Neural Networks for Robotics: An Engineering Perspective

Neural Networks for Robotics: An Engineering Perspective

Nancy Arana-Daniel
Alma Y. Alanis
Carlos Lopez-Franco

CRC Press
Taylor & Francis Group
Boca Raton London New York

CRC Press is an imprint of the
Taylor & Francis Group, an **informa** business

CRC Press
Taylor & Francis Group
6000 Broken Sound Parkway NW, Suite 300
Boca Raton, FL 33487-2742

© 2019 by Taylor & Francis Group, LLC
CRC Press is an imprint of Taylor & Francis Group, an Informa business

No claim to original U.S. Government works

ISBN 13: 978-0-367-73339-1 (pbk)
ISBN 13: 978-0-8153-7868-6 (hbk)

Version Date: 20180810

Library of Congress Cataloging-in-Publication Data

Names: Arana-Daniel, Nancy, author. | Lopez-Franco, Carlos, author. | Alanis, Alma Y., author.
Title: Neural networks for robotics : an engineering perspective / Nancy Arana-Daniel, Carlos Lopez-Franco, Alma Y. Alanis.
Description: Boca Raton, FL : CRC Press/Taylor & Francis Group, 2018. | "A CRC title, part of the Taylor & Francis imprint, a member of the Taylor & Francis Group, the academic division of T&F Informa plc." | Includes bibliographical references and index.
Identifiers: LCCN 2018017262| ISBN 9780815378686 (hardback : acid-free paper) | ISBN 9781351231794 (ebook)
Subjects: LCSH: Robots--Control systems. | Neural networks (Computer science)
Classification: LCC TJ211.35 .A73 2018 | DDC 629.8/92632--dc23
LC record available at https://lccn.loc.gov/2018017262

Visit the Taylor & Francis Web site at
http://www.taylorandfrancis.com

and the CRC Press Web site at
http://www.crcpress.com

Nancy Arana-Daniel dedicates this book to her husband, Angel, and her children, Ana, Sara and Angel, as well as her parents, Maria y Trinidad, and her brothers and sisters, Rodolfo, Claudia, Nora, Carlos, Ernesto, Gerardo and Paola.

Alma Y. Alanis dedicates this book to her husband, Gilberto, her mother, Yolanda, and her children, Alma Sofia and Daniela Monserrat.

Carlos Lopez-Franco dedicates this book to his wife, Paty, and his children, Carlos Alejandro, Fernando Yhael and Íker Mateo

Contents

Preface

Artificial Neural Networks (ANN) are among the most successful and popular methodologies of Artificial Intelligence to solve complex engineering problems due to their bio-inspired learning capabilities, which make them powerful and highly adaptable algorithms to design applications to achieve autonomy for robots.

There are several books devoted to presenting applications of ANN for robotics; most of them have become classic books due to them being presented in the 90s. Since then there have been many advances in neural networks and robotics. Furthermore, most of these classic books were specialized in robot control theory.

So this book offers the reader the opportunity to learn ANN for robotics applications with real-time implementations. All of the proposed methodologies include real-life scenarios to implement a wide range of ANN architectures to solve different kinds of robotic problems, ranging from mapping of rough environments, to object recognition (pattern classification) to problems of autonomous robot navigation. Our methodologies can be applied to different kinds of mobile robots: indoor, outdoor, holonomic, non-holonomic, land or aerial. Through the chapters of this book, the reader can find different methodologies to solving each stage of autonomous navigation for robots: from algorithms that solve the robot perception problem by recognizing objects and mapping environments and covering methods to plan paths, to algorithms that control holonomic, non-holonomic and aerial robots. Furthermore, an example of a robust navigation system which integrates path planning and low-level control of robots is shown.

Besides the real-time implementations, the book contains all the theory required to use the proposed methodologies for different applications beyond robotics systems. This book includes integration of different kinds of methodologies in order to solve problems typically found in autonomous navigation; therefore it is necessary to integrate computing, communications, control, pattern recognition and planning among others. This is the part of the so-called cyber-physical systems.

This book is organized as follows: Chapter 1 presents the application of the control theory system idea of dynamic system identification to solve the problem of rough terrain cost mapping. Using a Recurrent High Order Neural Network (RHONN), the cost function that represents the difficulty for traversing a patch of rough terrain is identified, and then, traversability cost maps of outdoors terrain are computed. These maps are employed by path planners to find optimal trajectories or to explore unknown terrain. Chapter 2 includes the design and implementation of geometric machine learning algorithms called Geometric Neural Networks (GNN) such as Clifford Support Vector Machines (CSVM), Quaternion SVM (QSVM) and hyper-ellipsoidal neuron for object recognition through images and pattern classification. These learning algorithms are designed using the mathematical framework of geometric algebra, and this allows them to accept input data that lie in complex and hypercomplex spaces and/or that represent geometric entities. Also, CSVM and QSVM are able to do multiclassification and to define high-nonlinear classes boundaries as we show in the experimental results of the chapter. Chapter 3 presents the application of a neural inverse optimal controller for non-holonomic robots. First, a recurrent high order neural network trained with an extended Kalman filter is used to identify the mathematical model of non-holonomic robots. Then, based on the neural model, an inverse optimal controller is designed with the Lyapunov function-based methodology. Finally, in order to show the applicability of the proposed identifier-controller scheme, it is applied to two non-holonomic robots: a wheeled differential robot and a tracked robot, to achieve real-time trajectory tracking under the presence of external and internal disturbances. Chapter 4 includes the integration of a neural inverse optimal controller with path planning and navigation for a non-holonomic mobile robot for indoors. The proposed methodology integrates four stages: a Reinforcement Learning (RL), Simultaneous Localization and Mapping (SLAM), a neural identifier and an inverse optimal neural controller. This allows for an autonomous navigation system to face widely changing environments over time. Then the integration of planning that is called RL-SLAM-identifier-controller, is implemented and applied in real time using a differential wheeled robot. In order to show the effectiveness of the proposed scheme this chapter includes examples of autonomous navigation in different scenarios including unknown disturbances (internal and external) as well as unknown information delays.

Chapter 5 presents a discrete time neural control of an omnidirectional mobile robot with visual feedback. The approach consists of synthesizing a suitable controller for each subsystem. First, the dynamics of each subsystem is approximated by an identifier using a discrete-time recurrent high order neural network, trained with an extended Kalman filter algorithm. Then, based on this neural model, an inverse optimal controller is synthesized to avoid solving the Hamilton Jacobi Bellman (HJB) equation. The desired trajectory of the robot is computed during navigation using a camera sensor. Simulation results are presented to illustrate the effectiveness of the proposed control scheme.

Chapter 6 presents a neural network-based controller for an aerial vehicle (UAV), which is treated as a multirotor system. In this chapter, we present a neural network-based controller for multirotors. The advantage of this controller over conventional PID is that the neural network can tune the PID even if the multirotor changes its parameters during the flight, and it is not necessary to linearize the UAV model.

This book can be used for self-learning as well as a textbook. The intended audience includes, but is not limited to, academia, industry, research engineers, graduate students and by people working in the areas of robotics, machine learning, deep learning, artificial intelligence, vision, modelling and control and others. However, due to artificial neural networks being a well-established research area with many applications in different scientific fields, ranging from engineering and science to social sciences to physics and economics, it is not possible to restrict the scope that this book can have regarding the possible applications of the approaches presented here.

Nancy Arana-Daniel
Alma Y. Alanis
Carlos Lopez-Franco

Guadalajara, Jalisco, México
August 2018

The authors thank CONACYT (Spanish acronym, which stands for National Council of Sciences and Technology), Mexico, for financially supporting the following projects: CB-256769 and CB-256880. They also thank CUCEI-UDG (Spanish acronym for University Center of Exact Sciences and Engineering of the University of Guadalajara), Mexico, for the support provided to write this book.

Nancy Arana-Daniel would like to give special thanks to her colleagues Carlos Villaseñor Padilla and Julio Esteban Valdes Lopez, and extend her thanks to Gehova Lopez Gonzalez, Roberto Valencia Murillo, and Eduardo Bayro, who have contributed their work, inspirations, and enriching discussions to the realization of this book.

Alma Y. Alanis also gives thanks for the support given by Fundacion Marcos Moshinsky. In addition, gratitude is extended to Edgar N. Sanchez, Jorge D. Rios and Edgar Guevara-Reyes, who have contributed in different ways to the composition of this book.

Carlos Lopez-Franco thanks Jose de Jesus Hernandez and Javier Gomez.

Abbreviations

ANN	Artificial Neural Networks
NN	Neural Networks
HONN	High Order Neural Networks
RHONN	Recurrent High Order Neural Networks
EKF	Extended Kalman Filter
LEARCH	**Lea**rning to se**arch** approach
SVM	Support Vector Machine
BPTT	Backpropagation Through Time
NNOR	Neural Network Object Recognizer
GA	Geometric Algebra
GNN	Geometric Neural Networks
CA	Clifford Algebra
CGA	Conformal Geometric Algebra
HCGA	Hyperconformal Geometric Algebra
CSVM	Clifford Support Vector Machines
QPSK	Quadrature Phase Shift Keying
QVSVC	Quaternion Valued Support Vector Classifier
CN	Conformal Neuron
HN	Hyperellipsoidal Neuron
HJB	Hamilton-Jacobi-Bellman
LCF	Lyapunov Control Function
IONC	Inverse Optimal Neural Controller
RMSE	Root Mean Square Error
SLAM	Simultaneous Localization and Mapping
ISRR	International Symposium on Robotics Research
ICRA	International Conference on Robotics and Automation IEEE
RL	Reinforcement Learning
DC	Direct Current Motor
UAV	Unmanned Aerial Vehicle
DOF	Degrees Of Freedom
PID	Proportional Integral Derivative Controller
AAD	Average Absolute Deviation
IBVS	Image-Based Visual Servo
VTOL	Vertical Take-Off and Landing
IMU	Inertial Measurement Unit
PD	Proportional Derivative

1

Recurrent High Order Neural Networks for Rough Terrain Cost Mapping

CONTENTS

1.1 Introduction

Autonomous robot navigation can be defined as the capability for a robot to move itself from one place to another without the helping of a human, i.e., without needing a human operator to achieve it. In order for a robot to achieve autonomous navigation, the next algorithms and representations are usually:

- A map of the environment in which the robot is going to navigate. A virtual representation or abstraction of the environment is required, and it has to be obtained using sensors such as cameras, laser, infrared, sonars, etc. Then these sensor readings have to be translated into a representation that the robot can use to make decisions about the actions that it has to make to navigate in the environment.

- A method to localize the robot in the environment and into the representation of it (in the map). With this process, the current position of the robot within the map is calculated.

- A path-planner. This is the algorithm that obtains the trajectory that the robot should follow in the environment to achieve a goal position in the map. It is usual that the obtained path is represented as a sequence of spatial coordinates regarding the map of the environment.

- A movement-planner. This algorithm obtains paths represented as a sequence of movements described by linear and angular velocities at which the robot should move to follow the desired trajectory.

- A controller. This is the method that computes the control actions (usually values of voltages and torques) that the robot has to perform to follow the desired trajectory.

As the reader will note, all the algorithms described above depend on having a good representation of the environment in which the robot will navigate. That is to say, it is very important to get a map that is accurate, with enough information and highly descriptive of the environment in order for the rest of the algorithms involved in autonomous navigation to obtain good results. For example, for a path planner algorithm to get an optimal path from a start to a goal position on the environment it is necessary to give it information about areas that it can traverse and places that are occupied by obstacles. Or even better, we can provide the path planner information about how easy (or how difficult) is to traverse a patch of terrain.

The types of environments in which a robot can navigate could be classified into structured and unstructured environments. A structured environment is an office-like environment (see Fig. 1.1), and one of the most used representations in robot navigation of this type of environment is the occupation grid. In occupation grids, each environment is divided using an $m \times n$ grid in which each cell is represented using one of two possible states: an occupied area or a free-obstacle area that can be traversed. This kind of environment can be found in indoors areas, where the furniture and walls are obstacles and the free floor is traversable terrain. On the other hand, non-structured environment or rough terrain (right side of Fig. 1.1) can be seen in outdoors, environments such as valleys, forests, mountains, rivers, roads, etc. In non-structured

FIGURE 1.1: Structured environment (left) and unstructured, rough environment (right).

terrain, it is not possible to define states as traversable and non-traversable (occupied or free cell of the grid), but instead, states have to be defined using a continuous range of values between these two states. The value of a grid cell or a state of rough terrain is used to represent the difficulty that a navigation agent has to traverse this area of the environment, i.e. its traversability cost.

In this chapter, we present a methodology to learn maps of rough environments that represent them through their traversability cost. We described non-structured terrain using four features of terrain for each cell of a grid: slope, the density of vegetation, debris and presence of water. Then a Recurrent High Order Neural Network (RHONN) is trained using these features as input data to learn a mapping from them into a traversability cost of the terrain. The approach presented in this chapter is based on two main ideas: 1) The dynamic system identification theory and 2) the inverse optimal control idea. So, we deal with dynamic outdoor environments as if we were dealing with a dynamic system, and then we use identifiers such as RHONNs to obtain the model of the system, i.e. the set of weights of the RHONN that map from input data to the output value. The optimal control principle is employed in the training methodology of the RHONN; as we do not know the cost function that optimally relates features of terrain with traversability costs, we instead have an expert behavior from which an agent (the RHONN) can learn the cost function from expert demonstrations. In this way, we obtain an identification RHONN that learns from expert demonstrations to map non-structured terrain to traverse costs.

1.1.1   Mapping background

In this section, we are going to present a quick literature review of mapping environments. The references cited are those that deal with the cost mapping problem.

The LEArning to seaRCH approach (LEARCH) [106] is a family of algorithms which employ the learning by demonstration methodology. First, a human expert traces a path that is considered as the optimal path from a start to a goal state on the environment; in order for a path planner to compute the same path, it is needed that in the cost map the path traced by the human expert is optimum. In this way, a search for a cost function is made that generates a cost map with these restrictions. This approach does not need apriori knowledge of the environment, and it can produce the whole cost map using the examples paths traced by an expert. Nevertheless, the cost of the states that are far away from those that belong to the path traced by the expert are very imprecise; they have a lot of error. Furthermore, if the robot deals with a very different environment than to those that it previously has seen, it must be trained again, i.e., it is not possible to re-use knowledge previously learned in new navigation episodes. The training stage has to be performed off-line because the learning systems of the approach are very complex and slow.

In [91] the computation of a probabilistic cost map for path planning in outdoor environments is proposed, where each one of the cells of the grid is represented by a probability distribution of the cost of the cell instead of just one scalar value. This approach allows for heuristic algorithms such as A^* to work better, consuming less time and memory. On the other hand, probabilistic cost map generation lacks an automated methodology to set the values of each cell of the environment grid, therefore it is necessary that an expert sets these values. Another disadvantage is that the knowledge obtained generating one cost map can not be generalized or reused.

Authors Kim, Sun, et al. in [64]defined a method for graduating the traversability cost of the terrain using an on-line non-supervised learning algorithm to classify images obtained using a stereoscopic camera and the previous experiences of the navigating robot agent. This is a highly adaptable algorithm; it can deal with new scenarios by comparing them with previously seen images. A big drawback is that the images used as a reference have to be stored to use them to compare with new images, and the amount of memory grows with respect to them.

In [112] a cost map is used that is generated using the visual features of the road. They train a Support Vector Machine (SVM) to weight the pixels of the sensed image and to determine if they belong to a valid road. This approach begins using a reduced database with data provided by a human expert and expands it with its experiences acquired in an on-line training. The main limitation of this methodology is that it is used only for an immediate local planning.

In [128] a Bayesian algorithm is presented to predict the cost of the terrain. This algorithm predicts the cost of the terrain using a generalization of the cost of traverse generated by several systems that are integrated into the robot. For this reason, human intervention is not necessary in the training stage, because the learning is supervised while the robot is navigating. The main restriction of this approach is that it totally depends on the subsystems (sensors) that the robot has.

Rebula et al. [107] train a little dog robot to classify the cost to traverse a patch of land; where each patch is modeled using a paraboloid. The dog classifies each patch that it traverses as good or bad depending on the distance that its legs slide. Then, using the database that stores the patches that the little dog classified, the algorithm determines the cost of a new patch of terrain depending on the paraboloids parameters. The disadvantage is that the classified patches are stored in a database without using any methodology to generalize the learned knowledge. Therefore the approach only classifies patches that are very similar to those stored in the database.

As we mentioned above, in this work a cost map methodology is presented that deals with dynamic unstructured environments using the dynamic system identification and inverse optimal control theories as frameworks. These theories are combined with the use of an intelligent identifier, named Recurrent High Order Neural Network. The RHONNs as they are described in [115, 119] are very efficient methods for modeling and identifying dynamic systems that can be linear and nonlinear. The environment in which the robot navigates can change because there are other navigating agents that cross the path; the weather can modify the features of the terrain, changing streams into rivers, causing landslides, producing debris, etc. So, taking into consideration the above, every rough outdoors environment can be considered as a dynamic system and therefore we are going to identify its dynamic model using RHONNs. Furthermore, the RHONNs have proven to be identifying methods that do not require a priori information about the model of the system, and they can adjust their parameters to add this information even in the presence of chaos [40]. The RHONNs can be trained in real time by adjusting their weights at the same time that the system to be identified is working.

1.2 Recurrent High Order Neural Networks, RHONN

The RHONNs are neural networks in discrete time that have been used to model and to identify dynamic systems in control problems [119]. They can model some dynamic systems, particularly if they have enough high order connections, then there is a weight vector such as the behavior of the RHONN that approximates any dynamic system for which the trajectory of the state remains in a compact set [115].

Recurrent high order neural networks are dynamic neural networks which are capable of capturing the dynamic response of complex non-linear systems [115, 120] and also systems with time-delay [5] due to characteristics like a flexible model that allows us to incorporate *a priori* information about the system to be identified, approximation capabilities, robustness against noise, on-line training and their dynamical behavior that is the result of their recurrent connections that improve their learning capabilities and performance [115, 120].

RHONN networks are the result of including high order interactions represented by triplets $(y_i y_j y_k)$, quadruplets $(y_i y_j y_k y_l)$, etc. to the first-order Hopfield model [115, 120].

The RHONN model used in this work is the series-parallel [108, 120] which is defined as:

$$\widehat{\chi}_i(k+1) = \omega_i^\top z_i(x(k), u(k)), \qquad i = 1, \cdots, n \tag{1.1}$$

with

$$z_i(x(k), u(k)) = \begin{bmatrix} z_{i_1} \\ z_{i_2} \\ \vdots \\ z_{i_{L_i}} \end{bmatrix} = \begin{bmatrix} \Pi_{j \in I_1} \xi_{i_j}^{d_{ij}(1)} \\ \Pi_{j \in I_2} \xi_{i_j}^{d_{ij}(2)} \\ \vdots \\ \Pi_{j \in I_{L_i}} \xi_{i_j}^{d_{ij}(L_i)} \end{bmatrix} \tag{1.2}$$

$$\xi_i = \begin{bmatrix} \xi_{i_1} \\ \vdots \\ \xi_{i_n} \\ \xi_{i_{n+1}} \\ \vdots \\ \xi_{i_{n+m}} \end{bmatrix} = \begin{bmatrix} S(x_1) \\ \vdots \\ S(x_n) \\ u_1 \\ \vdots \\ u_m \end{bmatrix} \tag{1.3}$$

$$S(\varsigma) = \frac{1}{1 + \exp(-\beta\varsigma)}, \quad \beta > 0 \tag{1.4}$$

where n is the state dimension, $\widehat{\chi}$ is state vector of the neural network, ω is the weight vector x is the plant state vector, and $u = [u_1, u_2, \ldots, u_m]^\top$ is the input vector to the neural network.

1.2.1 RHONN order

The model of the RHONN has the advantage of being adjustable because if we know the model of the system that we want to identify using the RHONN, we can use this information in the design of the architecture of the network by using the order of the system as the order of the RHONN. We can express the model of the RHONN in Eq. 1.6 as a linear combination:

$$x_i(k+1) = \omega_{i_1} \prod_{j \in I_1} \xi_{i_j}^{d_{ij}(1)} + \omega_{i_2} \prod_{j \in I_2} \xi_{i_j}^{d_{ij}(2)} + \cdots + \omega_{i_{Li}} \prod_{j \in I_{L_i}} \xi_{i_j}^{d_{ij}(L_i)} \tag{1.5}$$

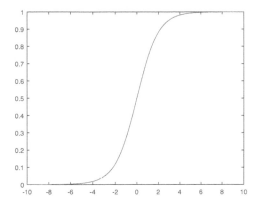

FIGURE 1.2: Logsig function.

And we assume that $d_{i_j}(\beta) = 1$ con $\beta - 1 \cdots L_i$ to simplify the equation; then we can rewrite the products as:

$$x_i(k+1) = \omega_{i_1}\xi_{i_1} + \omega_{i_2}\xi_{i_2} + \cdots + \omega_{i_\gamma}\xi_{i_\gamma} + \tag{1.6}$$

$$\omega_{i_{1,1}}\xi_{i_1}^2 + \omega_{i_{1,2}}\xi_{i_1}\xi_{i_2} + \cdots + \omega_{i_{1,\gamma}}\xi_{i_1}\xi_{i_\gamma} + \omega_{i_{2,2}}\xi_{i_2}^2 + \omega_{i_{2,3}}\xi_{i_2}\xi_{i_3} + \{\cdots + \omega_{i_{2,},\xi_{i_2}\xi_{i_\gamma}} + \cdots + \omega_{i_{\gamma,\gamma}}\xi_{i_\gamma}^2 + \tag{1.7}$$

$$\omega_{i_{1,1,1}}\xi_{i_1}^3 + \omega_{i_{1,1,2}}\xi_{i_1}^2\xi_{i_2} + \cdots + \omega_{i_{1,1,\gamma}}\xi_{i_1}^2\xi_{i_\gamma} + \omega_{i_{1,2,2}}\xi_{i_1}\xi_{i_2}^2 + \omega_{i_{1,2,3}}\xi_{i_1}\xi_{i_2}\xi_{i_3} + \cdots +$$
$$\omega_{i_{1,\gamma,\gamma}}\xi_{i_1}\xi_{i_\gamma}^2 + \omega_{i_{2,2,2}}\xi_{i_2}^3 + \omega_{i_{2,2,3}}\xi_{i_2}^2\xi_{i_3} + \omega_{i_{2,2,\gamma}}\xi_{i_2}^2\xi_{i_\gamma} + \omega_{i_{2,3,3}}\xi_{i_2}\xi_{i_3}^2 + \omega_{i_{2,3,4}}\xi_{i_2}\xi_{i_3}\xi_{i_4} + \cdots |$$
$$\omega_{i_{2,\gamma,\gamma}}\xi_{i_2}\xi_{i_\gamma}^2 + \cdots + \omega_{i_{\gamma,\gamma,\gamma}}\xi_{i_\gamma}^3 +$$

$$\vdots$$

$$\tag{1.8}$$

$$+\omega_{i_{1,\ldots,1_N}}\xi_{i_1}^N + \omega_{i_{1,\ldots,1_{N-1},2}}\xi_{i_1}^{N-1}\xi_{i_2} + \cdots + \omega_{i_{1,\ldots,1_{N-1},\gamma}}\xi_{i_1}^{N-1}\xi_{i_\gamma} + \omega_{i_{1,\ldots,1_{N-2},2,2}}\xi_{i_1}^{N-2}\xi_{i_2}^2$$
$$+\omega_{i_{1,\ldots,1_{N-2},2,3}}\xi_{i_1}^{N-2}\xi_{i_2}\xi_{i_3} + \cdots + \omega_{i_{1,\gamma,\ldots,\gamma_{N-1}}}\xi_{i_1}\xi_{i_\gamma}^{N-1} +$$

$$\vdots$$

$$+\omega_{i_{\gamma,\ldots,\gamma_N}}\xi_{i_\gamma}^N$$

$$\tag{1.9}$$

with $\gamma = n + m$. On Eq. 1.6, we can see terms with an single factor se ξ_i; these represent the first order terms of the network. Each of the second order terms grouped in Eq. 1.7 consists of two factors ξ_i; these represent the correlation between two input data $\xi_{i_a}\xi_{i_b}$; meanwhile in Eq. 1.8 are grouped the third order factors (made up of three factors) $\xi_{i_a}\xi_{i_b}\xi_{i_c}$. If we continue adding N factors ξ_i to the terms, we get the N order terms of the network, as we can see in Eq. 1.9. The maximum amount of different high order terms that we can obtain for each order is proposed in [48], as:

$$\binom{\gamma + j - 1}{j} = \frac{(\gamma + j - 1)!}{j!(n-1)!}, 1 \leq j \leq N \tag{1.10}$$

where j represents the terms of order j.

1.2.2   Neural network training

Neural network training is a process by which the neural network learns a task, this training can be on-line or off-line [50, 54]. The most common training algorithms for static neural networks and dynamic neural networks are backpropagation and backpropagation through time learning (BPTT), respectively [50, 53, 54]. BPTT is an effective trainning method for RHONN, but nevertheless it is very slow and therefore it is not possible to use it when real-time training is required.

1.2.2.1   Kalman filter

The Kalman filter, estimates the state of a linear system with additive state and output white noise using a recursive solution in which each update of the state is estimated from the previous estimated state and the new input data [53, 120].

1.2.2.2   Kalman filter training

The Kalman filter for neural networks offers advantages such as reduction of the epoch number and number of required neurons, on-line and off-line training implementation, and improvement of learning convergence. They are also more computationally efficient compared to the most used back-propagation methods [53, 120]. Moreover, they have proven to be reliable and practical [54, 120].

The training goal is to find the optimal weight vector which minimizes the prediction error. Due to the fact that the neural network mapping is non-linear the extended Kalman Filter (EKF) is required.

1.2.2.3   Extended Kalman filter-based training algorithm, EKF

One effective and efficient method to train RHONNs is the EKF [119], because its convergence is fast, and unlike the traditional Kalman filter algorithm, EKF can deal with nonlinear systems.

EKF estimates the neural network weights which become the state, and the error between the measured output of the plant and the output of the neural network is considered as additive white noise [53, 120].

The EKF-based training algorithm [120] for RHONN series-parallel (1.1) is (1.11):

$$\omega_i(k+1) = \omega_i(k) + \eta_i K_i(k) e_i(k) \tag{1.11}$$
$$K_i(k) = P_i(k) H_i(k) M_i(k)$$
$$P_i(k+1) = P_i(k) - K_i(k) H_i^{\top}(k) P_i(k) + Q_i(k)$$
$$M_i(k) = [R_i(k) + H_i^{\top}(k) P_i(k) H_i(k)]^{-1}$$

with

$$e_i(k) = x_i(k) - \hat{\chi}_i(k) \tag{1.12}$$

$$H_{ij} = \left[\frac{\partial \hat{\chi}_i(k)}{\partial \omega_{ij}(k)} \right]^{\top} \tag{1.13}$$

$$i = 1 \cdots n$$

where $\omega_i \in \Re^{L_i}$ is the on-line adapted weight vector, $K_i \in \Re^{L_i}$ is the Kalman gain vector, $e_i \in \Re$ is the identification error, $P_i \in \Re^{L_i \times L_i}$ is the weight estimation error covariance matrix, χ_i is the i-th state variable of the neural network, $Q_i \in \Re^{L_i \times L_i}$ is the estimation noise covariance matrix, $R_i \in \Re$ is the error noise covariance matrix and $H_i \in \Re^{L_i}$ is a vector in which each entry H_{ij} is the derivative of the neural network state $(\hat{\chi}_i)$ with respect to one neural network weight (ω_{ij}) and it is given by (1.13). P_i and Q_i are initialized as diagonal matrices with entries $P_i(0)$ and $Q_i(0)$, respectively. It is important to remark that $H_i(k)$, $K_i(k)$ and $P_i(k)$ for the EKF are bounded [129].

The computation of H_{ij} is the derivative of the neural network state $\hat{\chi}_i$, with respect to one neural network weight w_{ij}, and it is derived as:

$$H_{ij}(k) = \left[\frac{\partial \hat{\chi}_i(k)}{\partial \omega_{ij}(k)} \right]_{\omega_i(k) = \hat{\omega}_i(k+1)}, \quad i = 1, \cdots, n \text{ y } j = 1, \cdots, L_i . \tag{1.14}$$

$$\frac{\partial \hat{\chi}_i(k)}{\partial \omega(k)} = \frac{\partial \hat{\chi}_i(k)}{\partial x(k)} \frac{\partial x(k)}{\partial \omega(k)} \tag{1.15}$$

$$\hat{\chi}_i(k+1) = \omega_i^{\top} z(x(k), u(k)), i = 1, \cdots, n \tag{1.16}$$

$$\hat{\chi}_i(k+1) = F(x(k), u(k), \omega(k)) \tag{1.17}$$

Using the chain rule:

$$\frac{\partial \hat{\chi}_i(k+1)}{\partial \omega(k)} = \frac{\partial F(x(k), u(k), \omega(k))}{\partial x(k)} \frac{\partial x(k)}{\partial \omega(k)} + \frac{\partial F(x(k), u(k), \omega(k))}{\partial u(k)} \frac{\partial u(k)}{\partial w(k)} + \frac{\partial F(x(k), u(k), \omega(k))}{\partial \omega(k)} \frac{\partial \omega(k)}{\partial \omega(k)} \tag{1.18}$$

with the factors $\frac{\partial u(k)}{\partial \omega(k)} = 0$ y $\frac{\partial w(k)}{\partial \omega(k)} = 1$, therefore:

$$\frac{\partial \hat{\chi}_i(k+1)}{\partial \omega(k)} = \frac{\partial F(x(k), u(k), \omega(k))}{\partial x(k)} \frac{\partial x(k)}{\partial \omega(k)} + \frac{\partial F(x(k), u(k), w(k))}{\partial \omega(k)} \quad (1.19)$$

Matrices P_i , Q_i y R_i can be initialized as zero matrices in their respective dimensions.

1.3   Experimental Results: Identification of Costs Maps Using RHONNs

For this section, we are going to define some important symbols: an environmental map can be seen as a matrix $\mathcal{E} \in \mathbb{R}^{m \times n}$. Each cell of the matrix $u_{ij} \in \mathcal{E}$ is described using a vector $u \in \mathcal{F}$ where $\mathcal{F} \subseteq \mathbb{R}^l$ is the feature space, l is the number of features of the environment used to describe each path of terrain in each cell of the matrix $\mathcal{E}$. For a path planner to compute an optimal path, a traversability cost map is required; then to compute the cost map, we need a cost function $C : \mathcal{F} \to t_c$, where the traverse cost $t_c \in \mathbb{R}$; in other words, a function that maps the features of the environment with the traverse cost of it.

To approximate the cost function C, we use the RHONN due to its capabilities to reuse and generalize the learned knowledge. We deal with the environment as if we are dealing with dynamic complex systems. The immense quantity of rough environments and features that describe them, make necessary the application of a learning algorithm that helps to avoid the situation for each variation of environment the system uses a different cost function or that the system has to be trained to learn a different cost function. So, the generalization capability of RHONNs helps to minimize these problems [119], because they allow that by using a small set of training samples the learned knowledge can be generalized to environments that are similar to those that were seen in previous navigation episodes. All the above is possible thanks to the fact that the RHONNs are based on the Hopfield neural networks that are known because they behave as associative memories.

We designed the next general methodology to test the capability of the RHONN to identify traversability cost functions of rough environments:

1. Each map of rough terrain is divided using a grid as can be seen in Figs. 1.21, 1.24, 1.28 and 1.31.

2. Each patch of terrain contained in each cell of the grid is described using four features of the terrain: slope, the quantity of debris, vegetation density, and presence of water. We consider that the robot agent is not amphibian, so, in our experiments, the presence of wa-

ter on the terrain means that the terrain is not traversable for the robot (example in Figs. 1.3, 1.9, 1.15, 1.22, 1.25, 1.29 and 1.32).

3. Using the features maps described in the above item a human expert produces the cost maps to train the RHONN. In other words, a human expert uses the four features and interprets them as a real value that represents the difficulty that a robot agent would have to traverse the patch of terrain in the cell (see first row of Figs. 1.5, 1.11, 1.17, 1.23, 1.27, 1.30 and 1.33)

4. The cost maps produced by the human expert are used to train the RHONN as desired output values, meanwhile the features maps are used as input data. The RHONNs are trained to obtain generalizations of the learned cost maps (see second file of Figs. 1.5,1.11,1.17, 1.23, 1.27, 1.30 and 1.33)

In our experiments, we used an RHONN with input $u = \{u_1, u_2, u_3, u_4\}$ $\in \mathbb{R}^4$ where each feature u_i represents slope, the quantity of debris, vegetation density, and presence of water (in that order). The output $\hat{y} = x \in \mathbb{R}$ where x is a vector with an unique state , and a second order architecture of the network with $\{I_1, I_2, \cdots, I_{20}\} = \{[0\ 1], [0\ 2], [0\ 3], [0\ 4], [0\ 5], [1\ 1], [1\ 2], [1\ 3], [1\ 4], [1\ 5], [2\ 2], [2\ 3], [2\ 4], [2\ 5], [3\ 3], [3\ 4], [3\ 5], [4\ 4], [4\ 5], [5\ 5]\}$. The order of the neural network was determined experimentally. We use the lowest order that could imitate with the smallest error the behavior of the human expert.

Then our experiments are divided into two stages:

1. Synthetic dynamic environments. First, we work with artificial (synthetic) data in order to test the capabilities of the RHONN to generalize the learned knowledge when the RHONN deals with dynamic environments.

2. Real maps. Second, we test the RHONN using real environment maps to test the RHONN capabilities of reusing learned knowledge when the RHONN deals with environments with similar features.

1.3.1   Synthetic dynamic environments

In this stage of the experiments, we are going to test the RHONN capabilities to deal with dynamic environments. To have control of the percentage of change that can occur in a dynamic environment we use synthetic data, i.e., feature maps produced and changed artificially. The particular methodology designed for this experiments is:

1. First we generate base feature maps with random features (Figs. 1.3, 1.9, 1.15) and then a human expert produces a base cost map to train the RHONN as was explained in our general methodology 1.3.

2. Then we gradually change the base feature maps and the base cost map. We increase the changes in the base map: initially, we start with 10% of cells changed, and we increase the changes using a 10% step on each test until we reach a total of 90% cells changed. We change each cell value adding a 30% of its original value. Using the same knowledge learned by the RHONN with the base cost maps, we test using the changed maps to simulate dynamic environments.

The Tables 1.1, 1.2 and 1.3 show the mean squared error of the results of the RHONN obtained by comparing the output of the neural network with the values of the base cost map produced by the human expert. It can be noted from these results that even when we increase the number of cells changed to 90%, the identification error does not grow proportionally. In Figs. 1.8, 1.14, and 1.20 it can be seen that even though the maps are very different with respect to the training ones the results obtained using the RHONNs are very similar to the desired outputs. The above is because the RHONN does not learn maps as images, but it learns cost map policies in the form of cost functions, and therefore the neural network can generalize the knowledge to very different environments than those that it saw in the training phase.

In the next subsections, we show the results of three synthetic dynamic environments. We first show the four features map grids, i.e., the grids in which random base values of slope, the quantity of debris, vegetation density, and presence of water are shown. Orthogonal and lateral views of these maps are presented for the reader to appreciate the difficulty of the artificial terrains described with these features; it is important to remember that these base maps are the input data for training the RHONN. After, we present the map that the human expert produces using the four features base maps and that is also the desired output for the RHONN. In the figures that show the human expert map is also shown the result obtained with the RHONN so the reader can compare the results. In addition, and following the particular methodology described in this section, we present a table that shows the SME per experiment in which we change a percentage of number of cells of each map and use the changed maps as test data in the experiment (without re-training, but using only the knowledge learned by the RHONN with the base maps). Changed maps of features, the human expert map and the result obtained using the RHONN for 90% of changed cells are also shown after the table of SME.

1.3.1.1   Synthetic dynamic random environment number 1

The base maps of the four features produced for this experiment are shown in Figs. 1.3 and 1.4 (orthogonal and lateral views respectively) . With this base features maps the four dimension vector to describe each cell is constructed and used as input data for training the RHONN.

Using the random base feature maps of Fig. 1.3 a human expert produces the cost map shown in the top of Fig. 1.5 and this map is used as desired

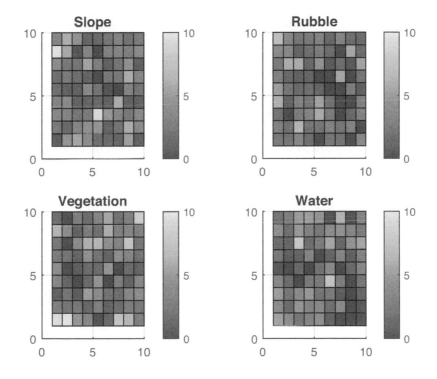

FIGURE 1.3: Random base features maps, orthogonal view.

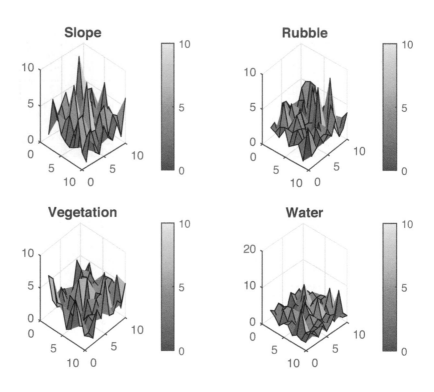

FIGURE 1.4: Lateral view of random base features maps on Fig. 1.3.

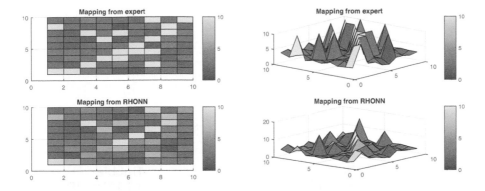

FIGURE 1.5: Human expert cost map (top) and result obtained using the RHONN (bottom) for random base features maps of Fig. 1.3.

TABLE 1.1: SME for test dynamic environments taking maps of Fig. 1.3 as base maps for training.

%of changed cells	SME
0%	0.9840
10%	1.3226
20%	2.8866
30%	2.7650
40%	3.0129
50%	2.1875
60%	3.6065
70%	2.5810
80%	7.2784
90%	3.7627

output for training the RHONN. The result obtained using the RHONN is shown in the bottom of Fig. 1.5

Now, using the learned knowledge acquired using the base maps of Fig. 1.3 we ran test experiments changing a number of cells corresponding to a different percentage of cells for each experiment to simulate dynamic changes in the environment. In Table 1.1 are shown the squared mean error for each experiment.

In Figs. 1.6 and 1.7 are shown maps of features with 90% of cells changed

Fig. 1.8 shows the human expert map and RHONN results obtained using the maps with 90% of changed cells.

At first glance, we can see the similarity between the cost maps of the human expert and that produced by the RHONN in Fig. 1.5. It is important to note that in Figs. 1.5 and 1.8 the human expert changes the costs of adjacent

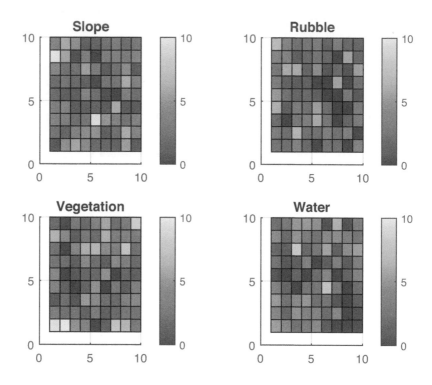

FIGURE 1.6: Map generated using as base map the one shown in Fig. 1.3 changing 90% of its cells.

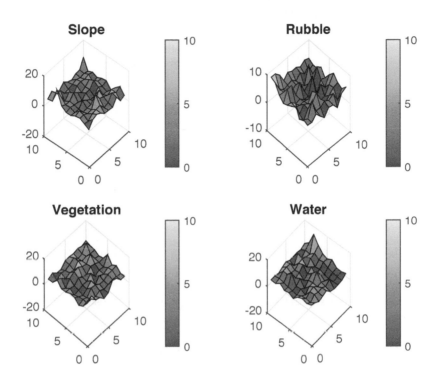

FIGURE 1.7: Lateral view of maps in Fig. 1.6.

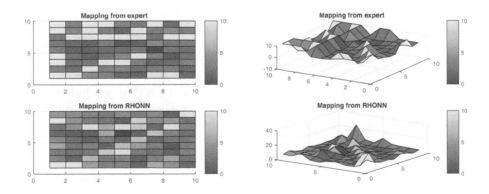

FIGURE 1.8: Human expert map and results using the RHONN with maps with 90% of changed cells.

TABLE 1.2: SME for test dynamic environments taking maps of Fig. 1.9 as base maps for training.

%of changed cells	SME
0%	1.1579
10%	1.3675
20%	1.4691
30%	1.7987
40%	1.5707
50%	2.2250
60%	2.5707
70%	2.2912
80%	2.0761
90%	4.2002

cells abruptly from a low value to a high value. Meanwhile in the RHONN these changes are softer, and this could help a navigation agent to know that it is approaching dangerous terrain to traverse.

In the next subsections, the order of figures is the same as in this subsection.

1.3.1.2   Synthetic dynamic random environment number 2

In this experiment, it can be seen in Fig. 1.11 that if we compare the cost map produced by the RHONN with the map of the human expert, in the cost map of the RHONN, again, the cells with higher values affected the value of the cells around them by increasing their values. As we mentioned in the previous subsection, this could help the robot as an alert that it is approaching such dangerous terrain as areas around water, cliffs, steep slopes, etc.

An interesting point to note occurs when we compare the SME table from 1.1 with the SME table of this experiment 1.2. In the last one the SME increased more slowly than in the previous experiment, but if we observe the cost maps in Figs. 1.11 and 1.14 that show the base map and the 90% changed map, we can appreciate a greater difference between them than between the maps of the previous experiment in Figs. 1.5 and 1.8. We conclude that the cost policy learned by the RHONN in this experiment is more robust (and therefore more generalizable) than the one learned in the previous experiment.

1.3.1.3   Synthetic dynamic random environment number 3

In this experiment the results are similar to the above. The RHONN approximated the result of the human expert (Fig. 1.15). In Table 1.3 the SME is increases gradually, but it goes down to the end. In Fig. 1.20 after changing 90% of the base maps cells, the map obtained using the RHONN is comparable to the one produced by the human expert.

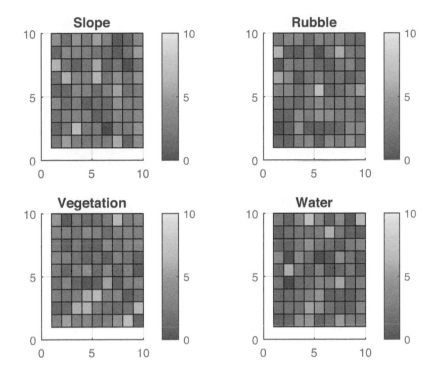

FIGURE 1.9: Random base features maps, orthogonal view.

TABLE 1.3: SME for test dynamic environments taking maps of Fig. 1.15 as base maps for training.

%of changed cells	SME
0%	1.0967
10%	1.2029
20%	1.7887
30%	2.6842
40%	3.0284
50%	1.5657
60%	3.6090
70%	3.7437
80%	4.0962
90%	1.8249

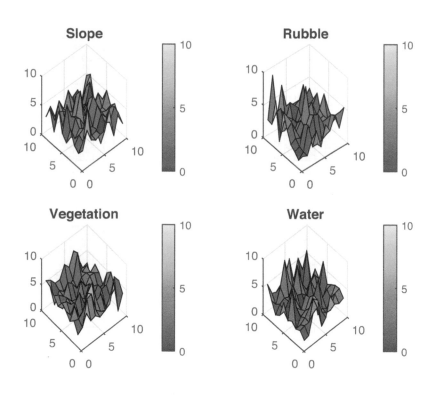

FIGURE 1.10: Lateral view of random base features maps on Fig. 1.9.

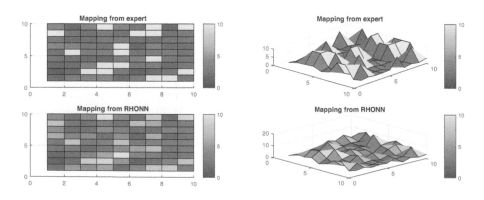

FIGURE 1.11: Human expert cost map (top) and result obtained using the RHONN (bottom) for random base features maps of Fig. 1.9.

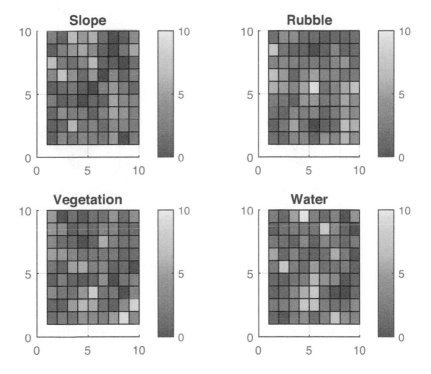

FIGURE 1.12: Map generated using as base map the one shown in Fig. 1.9 changing 90% of its cells.

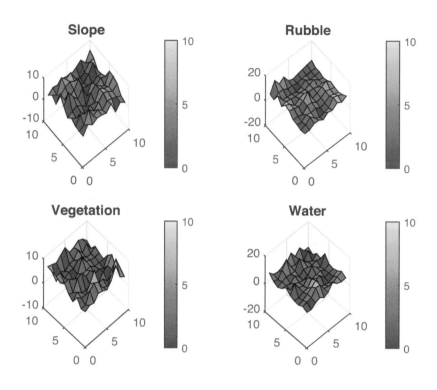

FIGURE 1.13: Lateral view of maps in Fig. 1.12.

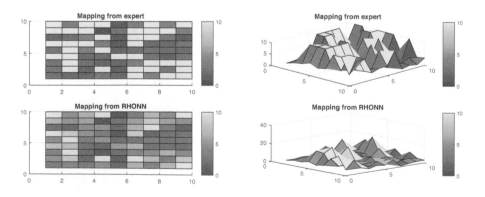

FIGURE 1.14: Human expert map and results using the RHONN with maps with 90% of changed cells for maps in Fig. 1.12.

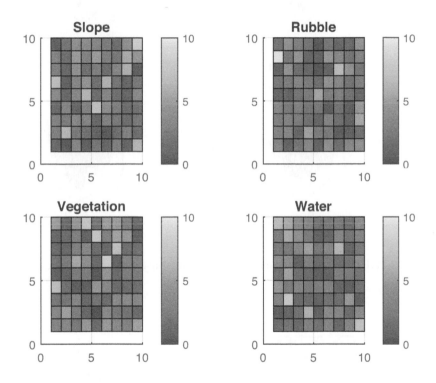

FIGURE 1.15: Random base features maps, orthogonal view.

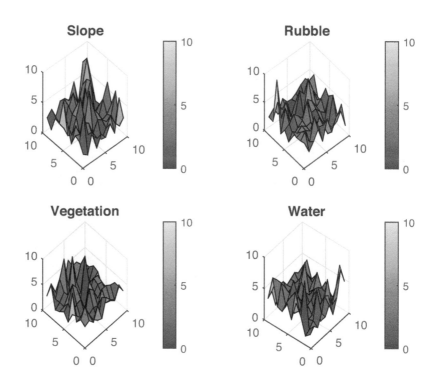

FIGURE 1.16: Lateral view of random base features maps on Fig. 1.15.

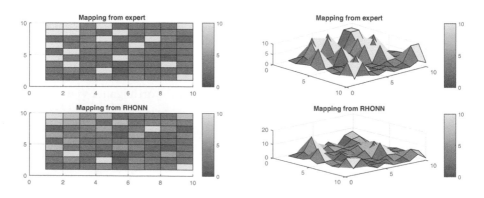

FIGURE 1.17: Human expert cost map (top) and result obtained using the RHONN (bottom) for random base features maps of Fig. 1.15.

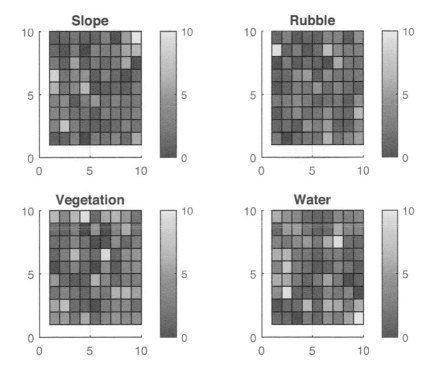

FIGURE 1.18: Map generated using as base map the one shown in Fig. 1.15 changing 90% of its cells.

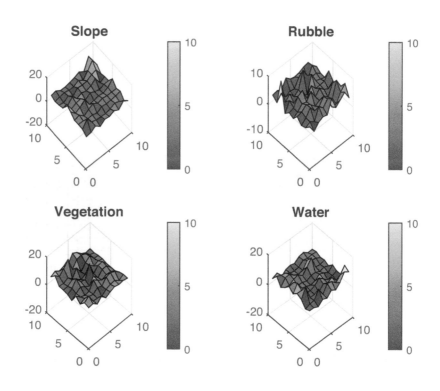

FIGURE 1.19: Lateral view of maps in Fig. 1.18.

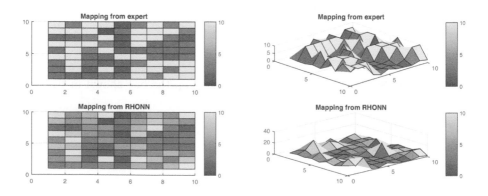

FIGURE 1.20: Human expert map and results using the RHONN with maps with 90% of changed cells for maps in Fig. 1.18.

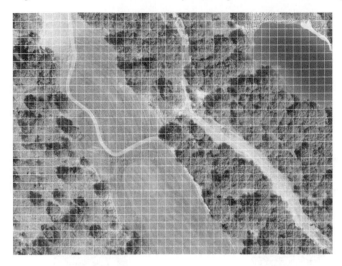

FIGURE 1.21: Satellital view of a grove.

1.3.2 Experiments using real terrain maps

We consider that the synthetic dynamic environments shown in Sec. 1.3.1 were more difficult experiments to test the RHONN capabilities to identify cost map functions because there is not a correlation between adjacent terrain as in the case of real terrain maps. So, in the experiments using real terrain maps we test that RHONNs are also an excellent choice to identify cost functions to typical outdoors environments in which an autonomous robot could navigate

1.3.2.1 Real terrain map: grove environment

In the first experiment using a real environment map we use a patch of terrain that contains a section of a grove with a small lake and a path that crosses the grove, see Fig. 1.21. We use the general methodology described in 1.3 and we split the map of the environment using a grid. Then, in Fig. 1.22 each cell in the grid was described using the four features that we already have used in synthetic maps to construct the input vector for the RHONN and to be used for the human expert to produce the cost map in the top of Fig. 1.23. It can be seen in the cost map that the RHONN produced (bottom of Fig. 1.23) that the neural network learned the cost function with high precision. It is important to note that the terrain with the presence of water was valued with high cost; meanwhile, cells of the grid with paths have the lowest values. This information helps a path planner to find lowest cost paths and to avoid dangerous terrain with water or dense vegetation where the robot can suffer damage.

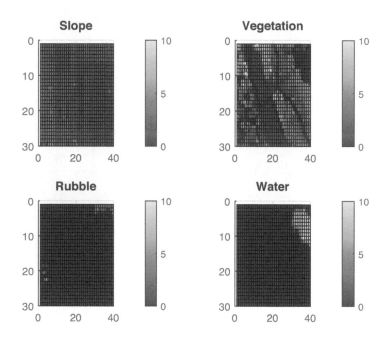

FIGURE 1.22: Feature maps for the grove in Fig. 1.21.

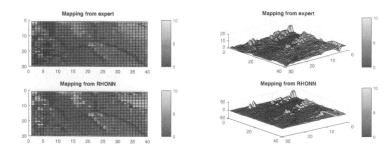

FIGURE 1.23: Human expert cost map (top) and result obtained using the RHONN (bottom) for the grove of Fig. 1.21.

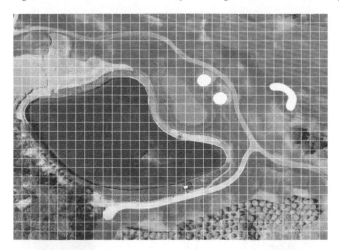

FIGURE 1.24: Satellite view of a golf course.

1.3.2.2 Real terrain map: golf course

In this experiment, we deal with a golf course Fig. 1.24. It can be seen that among its features it has a small lake, a grove and a large section of green area with a path that crosses it. If we obtain the describing features of each cell of the grid, we have the maps in Fig. 1.25.

In the map obtained by the RHONN in Fig. 1.27 it can be seen that the features of the environment with high-cost values include terrain with water and grove with low-cost values for paths. Comparing the result of the RHONN with the map produced by the human expert, we can note that they are similar. Nevertheless, in the RHONN map, we can see that on the edges of terrain with water and grove the RHONN has slightly increased the value of the cells (as happened on synthetic maps experiments). As we have mentioned the above is an advantage because this can give a robot a hint that dangerous terrain is near it. Furthermore, if we pay attention to the bottom left corner of the real map (a close up of this area is shown in Fig. 1.26), we can note that the map obtained by the RHONN has more details than the one produced by the human expert. This can help the path planner to obtain better trajectories when it is computing the path that the navigation agent will follow.

1.3.2.3 Real terrain map: forest

In the Fig. 1.28 we can see a forest with a path that crosses it. In Fig. 1.30 we can see described the paths with cells with low-cost meanwhile forest areas were valued with high-cost.

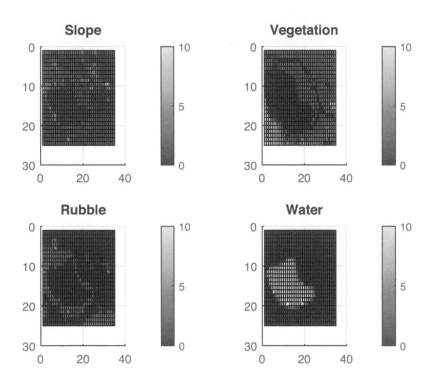

FIGURE 1.25: Feature maps of the golf course in Fig. 1.24.

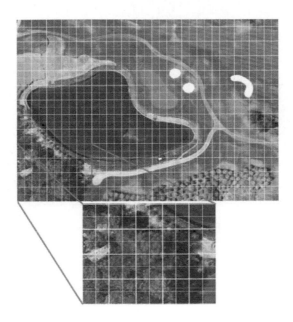

FIGURE 1.26: Bottom left corner close up of Fig. 1.24.

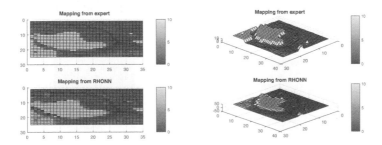

FIGURE 1.27: Human expert cost map (top) and result obtained using the RHONN (bottom) for the golf course of Fig. 1.24.

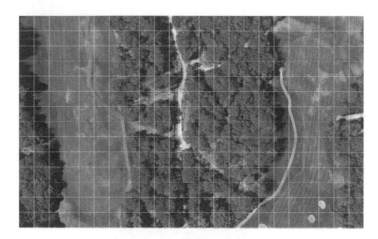

FIGURE 1.28: Satellite view of a forest.

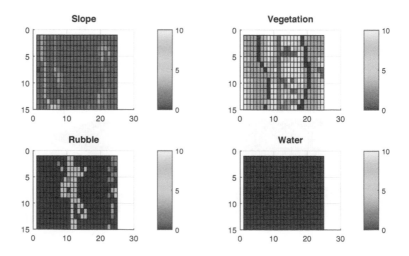

FIGURE 1.29: Feature maps of the forest of Fig. 1.28.

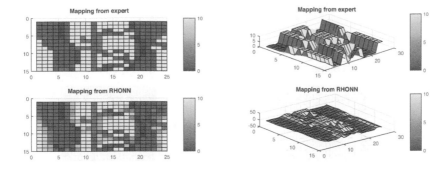

FIGURE 1.30: Human expert cost map (top) and result obtained using the RHONN (bottom) for the forest of Fig. 1.28.

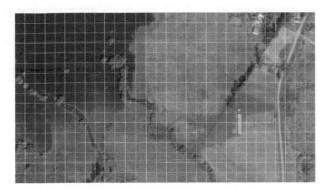

FIGURE 1.31: Satellite view of a rural area.

1.3.2.4 Real terrain map: rural area

In the last experiment using a real terrain map, we deal with a rural area with a water section and a path (Fig. 1.31). In the cost map computed by the RHONN, we can mark out the high-cost values for water and again, low-cost values for the path.

1.4 Conclusions

In this chapter, we have tested the traversability cost identification capabilities of the RHONN when we are dealing with dynamic environments. We proved that the RHONN is an excellent choice to recognize the dynamic model of rough outdoors terrains, even when we worked using maps produced by a human expert without the need of adding complexity to the RHONN model.

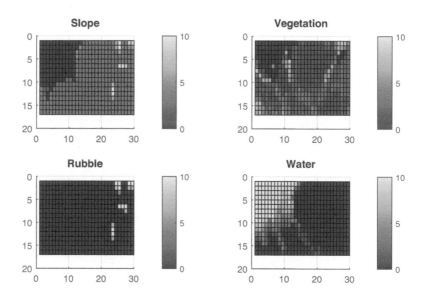

FIGURE 1.32: Feature maps of the rural area of Fig. 1.31.

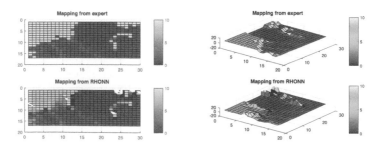

FIGURE 1.33: Human expert cost map (top) and result obtained using the RHONN (bottom) for the rural area of Fig. 1.31.

Furthermore, we can see for the experimental results that the RHONN learned the cost function of an environment even when this had been suffering changes. The above is very helpful in cases when we are dealing with real terrain, and it can suffer landslides, floods, earthquakes, etc. that can change the features of the environment drastically. These good results areas were obtained thanks to the capabilities of the RHONNs to identify and model complex dynamic systems.

2

Geometric Neural Networks for Object Recognition

CONTENTS

2.1   Object Recognition and Geometric Representations of Objects

We can define the object recognition task as the process that consists of distinguishing the common features of a type of objects to assign them a label that describes them. That is, we can say that the object recognition process is a classification task that collects similar objects into a set, and separates them from those objects that are different. So, object recognition is an important task in robotics; in order for a robot to interact with its environment, it has to know it, i.e., it has to recognize the objects that surround it.

For example in Chapter 1, a cost mapping for rough environments methodology was presented; there, we assume that the robot has sensors to read environmental features, such as: slope, the density of vegetation, the presence of

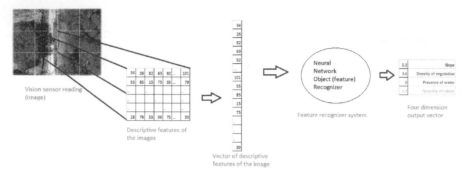

FIGURE 2.1: Process of environmental feature recognition.

water and the quantity of rubble. The first feature can be measured using two positions of the visible environment, but the rest of them imply that the robot has a recognition system that helps it relate the environmental characteristic (or the sensor reading that describes it) with a scalar value.

The above is equivalent to the robot performing a classification task, in which the input data to the classifier is a description vector of the environment that the robot is sensing and the output vector represents the quantity of the classes of features that the robot is recognizing. For example, the robot can see a patch of terrain like the one shown to the left of Fig. 2.1; then using a grid to split the terrain into twelve cells and picking one of the cells of the center we can take the matrix of feature values of the image (for example, gray values, edges, SURF or SIFT features, [13], [84] etc.) in the cell and use them as the input vector to a neural network classifier (or neural network object recognizer, NNOR). Then, we can get an output vector of dimension four, where each value represents the quantity of the four features that the NNOR recognizes in the image, see Fig. 2.1, to automate the creation of maps for training the system presented in Chapter 1.

In the same way, for a robot to be able to manipulate objects, it has to recognize them in order to know which object it has to pick up. So, the robot has to recognize objects and their poses, and it has to classify them according to the kind of grip with which it has to manipulate them, see Figs. 2.2. So these examples illustrate the importance of a recognition object system in order to achieve real robot autonomy. In this chapter, we are going to give a brief survey of three important approaches for object recognition using geometric neural networks that the authors have presented before. These are based on frameworks of Geometric Algebra (GA) [55, 103] and neural networks; we called them Geometric Neural Networks (GNN). These methods take advantage of the great power of representation of geometric entities and their operators that the GA allows (see next section), to design learning methods that could accept input data that represent geometric entities. Be-

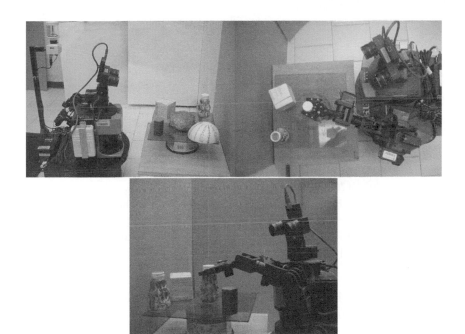

FIGURE 2.2: Robot sensing the environment. Robot recognizing and picking up the selected object (orthogonal and lateral views).

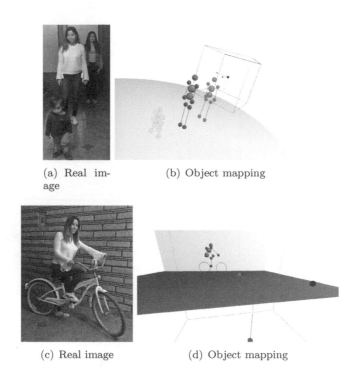

(a) Real image

(b) Object mapping

(c) Real image

(d) Object mapping

FIGURE 2.3: Object mapping of human bodies using geometric entities: spheres, lines, and circles.

sides the above, the presented GNNs use geometric algebra in the design of their learning algorithm to do multiclassification and/or to be able to define high-nonlinear classes boundaries as we show in the next sections.

2.1.1 Geometric representations and descriptors of real objects

The real world has a lot of objects with shapes that can be described and modeled or reconstructed using regular geometric entities, as well as a lot of free-form shape objects. The free-form shape objects are those that cannot be approximated using multiples planes [22]. Taking into consideration the first-mentioned type of objects, we can use the object mapping approach idea of representing objects using a set of geometric entities to get compact and rich-information maps of environments; i.e., maps that with few parameters can illustrate enough information about the structure of the environment and the objects inside it, see Figs. 2.3.

For example, in [92], an approach was present to represent rough terrain

FIGURE 2.4: Environmental object mapping: we are representing debris using points, slopes using planes and trees using spheres.

features through conformal geometric entities in order to give a robot enough information about the environment in which it was navigating, and in this way, a path planner could compute an optimal trajectory in this kind of difficult-to-explore terrain. In Figs. 2.4 we can see three kinds of rough terrain features that have been represented using spheres, planes, and points.

Then, using descriptors that are codified regarding geometric entities, we can use compact representations of objects to recognize (using a neural network classifier), model and reconstruct them. In the next section, we are going to introduce the mathematical framework used to codify the geometric entities in the form of multivectors.

2.2 Geometric Algebra: An Overview

Geometric Algebra also known as Clifford Algebra (CA), in fact represents a family of algebras that depend on a chosen vector space and on a special kind of product (the geometric product). The GA is a mathematical framework where we can find embedded concepts from linear algebra, tensor algebra, quaternion algebra, complex numbers and others. We represent a geometric algebra with $G_{p,q,r}$ [1].

Let G_n denote the geometric algebra of n-dimensions. This is a graded linear space. As well as vector addition and scalar multiplication we have a non-commutative product which is associative and distributive over addition – the *geometric* or *Clifford product*. A further distinguishing feature of the algebra is that any vector squares to give a scalar. Next we define the geometric product of two vectors a and b as a sum of their inner product (symmetric part) and their wedge product (antisymmetric part)

$$ab = a \cdot b + a \wedge b, \tag{2.1}$$

where the inner product $a \cdot b$ and the outer product $a \wedge b$ are further defined by

$$\begin{aligned} a \cdot b &= \tfrac{1}{2}(ab + ba) \\ a \wedge b &= \tfrac{1}{2}(ab - ba). \end{aligned} \tag{2.2}$$

The inner product of two vectors is the standard *scalar* or *dot* product and produces a scalar. The outer or wedge product of two vectors is a quantity known in Clifford algebras as a *bivector*. We think of a bivector as an oriented area in the plane containing a and b, formed by sweeping a along b. Thus, $b \wedge a$ will have the opposite orientation making the wedge product anti-commutative as given in (2.2). The outer product is immediately generalizable to higher dimensions – for example, $(a \wedge b) \wedge c$, a *trivector*, is interpreted as the oriented volume formed by sweeping the area $a \wedge b$ along vector c. The outer product of k vectors is a k-vector or k-blade, and such a quantity is said to have *grade* k. A multivector $A \in G_n$ is the sum of k-blades of different or equal grade. The letter $A = \langle A \rangle_r$ stands for an *homogeneous* blade of grade r.

2.2.1   The geometric algebra of n-D space

In an n-Dimensional (nD) space V^n we can introduce an orthonormal basis of vectors $\{e_i\}$, $i = 1, ..., n$, such that $e_i \cdot e_j = \delta_{ij}$. This leads to a basis

$$1, \quad \{e_i\}, \quad \{e_i \wedge e_j\}, \quad \{e_i \wedge e_j \wedge e_k\}, ...,$$
$$e_1 \wedge e_2 \wedge ... \wedge e_n = I, \tag{2.3}$$

which spans the entire geometric algebra G_n. Here I is the hypervolume called pseudoscalar which commutes with all the multivectors and it is used as a dualization operator as well. Note that the basis vectors are not represented by bold symbols. Any multivector can be expressed in terms of this basis.

Because the addition of k-vectors (homogeneous vectors of grade k) is

---

[1]For a more extensive introduction to Geometric Algebra please see [55], [103].

closed and the multiplication of a k-vector by a scalar is another k-vector, the set of all k-vectors is a vector space, denoted $\overset{k}{\bigwedge} V^n$. Each of these spaces is spanned by $\begin{pmatrix} n \\ k \end{pmatrix}$ k-vectors, where $\begin{pmatrix} n \\ k \end{pmatrix} := \frac{n!}{(n-k)!k!}$. Thus, our geometric algebra G_n, which is spanned by $\sum_{k=0}^{n} \begin{pmatrix} n \\ k \end{pmatrix} = 2^n$ elements, is a direct sum of its homogeneous subspaces of grades 0, 1, 2, ..., n, that is,

$$G_n = \overset{0}{\bigwedge} V^n \oplus \overset{1}{\bigwedge} V^n \oplus \overset{2}{\bigwedge} V^n \oplus \cdots \oplus \overset{n}{\bigwedge} V^n \tag{2.4}$$

where $\overset{0}{\bigwedge} V^n = R$ is the set of real numbers and $\overset{1}{\bigwedge} V^n = V^n$ corresponds to the linear nD vector space. Thus, any multivector of G_n can be expressed in terms of the basis of these subspaces.

In this book we will specify a geometric algebra G_n of the n dimensional space by $G_{p,q,r}$, where p, q and r stand for the number of basis vectors which square to 1, -1 and 0 respectively and fulfill n=p+q+r. Its even subalgebra which has multivectors of even grade will be denoted by $G_{p,q,r}^+$. Note however that the set of odd multivectors is not a subalgebra of $G_{p,q,r}$.

In the nD space there are multivectors of grade 0 (scalars), grade 1 (vectors), grade 2 (bivectors), grade 3 (trivectors), etc... up to grade n. Any two such multivectors can be multiplied using the geometric product. Consider two multivectors A_r and B_s of grades r and s respectively. The geometric product of A_r and B_s can be written as

$$A_r B_s = \langle AB \rangle_{r+s} + \langle AB \rangle_{r+s-2} + \cdots + \langle AB \rangle_{|r-s|} \tag{2.5}$$

where $\langle M \rangle_t$ is used to denote the t-grade part of multivector M, e.g. consider the geometric product of two vectors $ab = \langle ab \rangle_0 + \langle ab \rangle_2 = a \cdot b + a \wedge b$. Another simple illustration is the geometric product of $A = 5e_3 + 3e_1e_2$ and $b = 9e_2 + 7e_3 \in G_{3,0,0}$:

$$
\begin{aligned}
Ab &= 35(e_3)^2 + 27e_1(e_2)^2 + 45e_3e_2 + 21e_1e_2e_3 \\
&= 35 + 27e_1 - 45e_2e_3 + 21I. \tag{2.6}
\end{aligned}
$$

Note here that for $e_i e_i = (e_i)^2 = e_i \cdot e_i = 1$ and $e_i e_j = e_i \wedge e_j$ the geometric product of equal unit basis vectors equals 1 and the product of different unit bases is equal to their wedge, which for simple notation can be omitted.

Using Eq. (2.5) we can express the inner and outer products for the multivectors as

$$
\begin{aligned}
A_r \cdot B_s &= \langle A_r B_s \rangle_{|r-s|} \tag{2.7} \\
A_r \wedge B_s &= \langle A_r B_s \rangle_{r+s}. \tag{2.8}
\end{aligned}
$$

In order to deal with more general multivectors, we define the **scalar product**

$$A * B = \langle AB \rangle_0 . \tag{2.9}$$

and for an r-grade multivector $A_r = \sum_{i=0}^{r} \langle A_r \rangle_i$, the following operations are defined:

$$\text{Grade Involution: } \widehat{A}_r = \sum_{i=0}^{r} (-1)^i \langle A \rangle_i, \tag{2.10}$$

$$\text{Reversion: } A_r^\dagger = \sum_{i=0}^{r} (-1)^{\frac{i(i-1)}{2}} \langle A \rangle_i, \tag{2.11}$$

$$\text{Clifford Conjugation: } \widetilde{A}_r = \widehat{A}_r^\dagger$$

$$= \sum_{i=0}^{r} (-1)^{\frac{i(i+1)}{2}} \langle A \rangle_i. \tag{2.12}$$

The **grade involution** simply negates the odd-grade blades of a multivector. The **reversion** can also be obtained by reversing the order of basis vectors making up the blades in a multivector and then rearranging them to their original order using the anti-commutativity of the Clifford product.

The scalar product $*$ is positive definite, i.e. one can associate with any multivector $A = \langle A \rangle_0 + \langle A \rangle_1 + \cdots + \langle A \rangle_n$ a unique positive scalar **magnitude** $|A|$ defined by

$$|A|^2 = A^\dagger * A = \sum_{r=0}^{n} |A_r|^2 \geq 0, \tag{2.13}$$

where $|A| = 0$ if and only if $A = 0$. For an homogeneous multivector A_r its *magnitude* is defined as $|A_r| \equiv \sqrt{A_r^\dagger A_r}$.

In particular, for an *r-vector* A_r of the form $A_r = a_1 \wedge a_2 \wedge \cdots \wedge a_r$: $A_r^\dagger = (a_1 \cdots a_{r-1} a_r)^\dagger = a_r a_{r-1} \cdots a_1$ and thus $A_r^\dagger A_r = a_1^2 a_2^2 \cdots a_r^2$, so, we will say that such a r-vector is null if and only if it has a null vector for a factor. If in such *factorization* of A_r p, q and s factors square in a positive number, negative and zero, respectively, we will say that A_r is a r-vector with signature (p, q, s). In particular, if $s = 0$ such a *non-singular* r-vector has a multiplicative inverse

$$A^{-1} = (-1)^q \frac{A^\dagger}{|A|^2} = \frac{A}{A^2} \tag{2.14}$$

In general, the *inverse* A^{-1} of a multivector A, if it exists, is defined by the equation $A^{-1}A = 1$.

2.2.2   The geometric algebra of 3-D space

The basis for the geometric algebra $G_{3,0,0}$ of the 3-D space has $2^3 = 8$ elements and is given by:

$$\underbrace{1}_{scalar} , \underbrace{e_1, e_2, e_3}_{vectors}, \underbrace{e_1 e_2, e_2 e_3, e_3 e_1}_{bivectors}, \underbrace{e_1 e_2 e_3 \equiv I}_{trivector}.$$

In $G_{3,0,0}$ a *typical* multivector v will be of the form $v = \alpha_0 + \alpha_1 e_1 + \alpha_2 e_2 + \alpha_3 e_3 + \alpha_4 e_2 e_3 + \alpha_5 e_3 e_1 + \alpha_6 e_1 e_2 + \alpha_7 I_3 = <v>_0 + <v>_1 + <v>_2 + <v>_3$, where the α_i's are real numbers and $<v>_0 = \alpha_0 \in \bigwedge^0 V^n$, $<v>_1 = \alpha_1 e_1 + \alpha_2 e_2 + \alpha_3 e_3 \in \bigwedge^1 V^n$, $<v>_2 = \alpha_4 e_2 e_3 + \alpha_5 e_3 e_1 + \alpha_6 e_1 e_2 \in \bigwedge^2 V^n$, $<v>_3 = \alpha_7 I_3 \in \bigwedge^3 V^n$.

In geometric algebra a rotor (short name for rotator), R, is an even-grade element of the algebra which satisfies $R\widetilde{R} = 1$, where $\widetilde{R}$ stands for the *conjugate* of R using Eq. (2.12).

If $\mathcal{A} = \{a_0, a_1, a_2, a_3\} \in \mathcal{G}_{3,0,0}$ represents a unit quaternion, then the rotor which performs the same rotation is simply given by

$$
R = \underbrace{a_0}_{scalar} + \underbrace{a_1(Ie_1) - a_2(Ie_2) + a_3(Ie_3)}_{bivectors}
$$

$$
= a_0 + a_1 e_2 e_3 - a_2 e_3 e_1 + a_3 e_1 e_2. \tag{2.15}
$$

The quaternion algebra is therefore seen to be a subset of the geometric algebra of 3-space. According to Eq. (2.12) for the 3D Euclidean geometric algebra the conjugate of the rotor is $\widetilde{R} = a_0 - a_1 e_2 e_3 + a_2 e_3 e_1 - a_3 e_1 e_2$.

The transformation in terms of a rotor $a \mapsto Ra\widetilde{R} = b$ is a very general way of handling rotations; it works for multivectors of any grade and in spaces of any dimension in contrast to quaternion calculus. Rotors combine in a straightforward manner; i.e. a rotor R_1 followed by a rotor R_2 is equivalent to a total rotor R where $R = R_2 R_1$.

2.2.3 Conformal geometric algebra

The Conformal Geometric Algebra (CGA) is a 5-D Geometric Algebra that uses a conformal transformation to span Euclidean geometric algebra represented by G_3, to $G_{4,1}$. One of the advantages of using more dimensions is that problems are intuitively and easily formulated.

In CGA we use the Euclidean basis vectors e_1, e_2, e_3 that we can span with a conformal transformation over the Minkowski's plane with the basis e_-, e_+ that holds the Eq. 2.16. We can define another basis called null with Eq. 2.17

$$
e_+^2 = -1, \quad e_-^2 = 1, \quad e_+ \cdot e_- = 0. \tag{2.16}
$$

$$
e_0 = \frac{1}{2}(e_- - e_+), \quad e_\infty = e_- + e_+. \tag{2.17}
$$

Where e_0 represents the 3D origin and e_∞ represents the point to the infinity, they are null vectors and hold the Eq. 2.18

$$
e_0^2 = e_\infty^2 = 0, \quad e_\infty \cdot e_0 = -1. \tag{2.18}
$$

With this basis, geometric entities are represented as multivectors; in fact,

TABLE 2.1: Geometric entities on CGA.

Entity	IPNS	OPNS
Point	$x_C = x_E + \frac{1}{2}x_E^2 e_\infty + e_0$	$x_C^* = s_1 \wedge s_2 \wedge s_3 \wedge s_4$
Sphere	$s = x_E - \frac{1}{2}(x_E - r)^2 e_\infty + e_0$	$s^* = x_1 \wedge x_2 \wedge x_3 \wedge x_4$
Plane	$P = n + d e_\infty$	$P^* = x_1 \wedge x_2 \wedge x_3 \wedge e_\infty$
Line	$L = n I_E - e_\infty m I_E$	$L^* = x_1 \wedge x_2 \wedge e_\infty$
Circle	$C = s_1 \wedge s_2$	$C^* = x_1 \wedge x_2 \wedge x_3$

there is a dual representation for every entity; the inner product null space representation (IPNS) and the outer product null space representation (OPNS).

The basic geometric entities and their representations are listed in Table 2.1; where x_E stand for an euclidean point, x_C for a conformal point, r for the radius of the sphere, n a unitary vector and d a distance from the origin (Hesse representation of the plane) and finally n and m stand for the Plucker coordinates of the line.

We can use the same basis to represent rigid transformations, in general any conformal transformation can be expressed as Eq. 2.19.

$$\sigma X' = H X \tilde{H} \tag{2.19}$$

where $X, X' \in G_{4,1}$, H is a versor and σ a scalar; σ is applied to ensure that the null cone is invariant under H.

Rotations are represented by *Rotors* and they are defined in the form of Eq. 2.20.

$$R = e^{-\frac{\theta}{2}n} \tag{2.20}$$

where n represents the dual entity of the rotation axis and θ represents the rotation angle. We also define the *translator* in the form of Eq. 2.21.

$$T = 1 + \frac{1}{2}t e_\infty \tag{2.21}$$

t is the Euclidean translation. The null cone is invariant under R and T so $\sigma = 1$; then the rigid transformations are represented by Eq. 2.22.

$$X' = R X \tilde{R}, \quad X' = T X \tilde{T}. \tag{2.22}$$

We can use the last equations to express a rigid body motion; we denote such an operator and the conformal transformation with Eq. 2.23.

$$M = T R, \quad X' = M X \tilde{M}. \tag{2.23}$$

This equation of rigid movement can be applied on every entity of the algebra.

2.2.4 Hyperconformal geometric algebra

The most used geometric algebra is perhaps the Conformal Geometric Algebra (CGA) for the 3D case $G_{4,1}$ that we presented in the above section, and its extension to n-dimensions $G_{n+1,1}$ to represent the $I\!R^n$ vector space. In this algebra, we can represent points, pair of points, lines, planes, circles and spheres. Many algorithms for robotics and machine vision [16] have been developed with this algebra, for example, we can find path planning algorithms [92], pose estimation [113], structure extraction from motion [10], geometric entities detection [14], [83] and robotic mapping algorithms [110], [111], among others.

However, for using more complex geometric entities as algebra elements, we have to use other algebras. The geometric algebra $G_{6,3}$ [144] is a generalization of $G_{4,1}$ and in this algebra we can represent as multivectors geometric entities such as ellipsoids, planes, pair of planes, pseudo-cylinders, spheres and others deformed quadratic surfaces. In [133], we extended $G_{6,3}$ for any dimension in the so-called Hyperconformal Geometric Algebra (HCGA) with a notation $G_{2n,n}$ for representing the vector space R^n, this algebra is constructed by using a homogeneous stereographic projection in (2.24) over every coordinate. In Figure 2.5, we show a graphical representation of the projection.

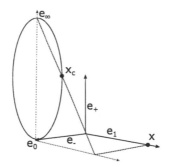

FIGURE 2.5: Homogeneous stereographic projection.

$$x_c = 2\frac{x}{x^2+1}e_1 + \frac{x^2-1}{x^2+1}e_+ + e_- \qquad (2.24)$$

Given a tridimensional space with basis (e_x, e_y, e_z), we can calculate the entities in $G_{6,3}$ using the null basis. For a point $x = p_x e_x + p_y e_y + p_z e_z$, we get (2.25), and for the ellipsoid with center (c_x, c_y, cz) and semiaxis (r_x, r_y, r_z) in (2.26).

$$X = p_x e_x + p_y e_y + p_z e_z + \frac{1}{2}(p_x^2 e_{\infty x} + p_y^2 e_{\infty x} + p_z^2 e_{\infty z}) + e_0 \quad (2.25)$$

$$E = \frac{c_x}{r_x^2}e_x + \frac{c_y}{r_y^2}e_y + \frac{c_z}{r_z^2}e_z + \frac{1}{2}\left(\frac{c_x^2}{r_x^2} + \frac{c_y^2}{r_y^2} + \frac{c_z^2}{r_z^2} - 1\right)e_\infty + \frac{1}{r_x^2}e_{0x} + \frac{1}{r_y^2}e_{0y} + \frac{1}{r_z^2}e_{0z}$$

$$(2.26)$$

Other geometric entities are derived from the ellipsoid, as we show in the Table 2.2.

TABLE 2.2: Geometric entities that can be represented by a deformed ellipsoid.

Entity	Representation
Sphere	$S = E$ if $r_x = r_y = r_z$
Pseudo-cylinder	$C = \lim_{r_i \to \infty} E$, $i \in \{x, y, z\}$
Pair of planes	$C = \lim_{r_i, r_j \to \infty} E$, $i, j \in \{x, y, z\}$ and $i \neq j$

2.2.5   Generalization of $G_{6,3}$ into $G_{2n,n}$

To generalize the idea of Zamora in [144] to R^n we can use the same homogeneous projection of Fig. 2.5 and Eq. 2.24 to map each n- real dimension into $R^{2n,n} = R^{2,1} \times R^{2,1} \times \cdots \times R^{2,1}$. The notation for $G_{2n,n}$ represents the projection of one vector in the direction of another using parenthesis, for example $e_{(+2)}$ denotes the projection of e_2 in the direction of e_+. So, using this extended notation, we can define the null bases of the algebra as:

$$e_{(\infty i)} = \qquad\qquad e_{(+i)} + e_{(-i)} \qquad\qquad (2.27)$$

$$e_{(0i)} = \qquad\qquad \frac{e_{(-i)} + e_{(+i)}}{2} \qquad\qquad (2.28)$$

$$e_\infty = \qquad\qquad \frac{1}{n}(e_{(\infty 1)} + e_{(\infty 2)} + \cdots + e_{(\infty n)}) \qquad\qquad (2.29)$$

$$e_0 = \qquad\qquad e_{(01)} + e_{(02)} + \cdots + e_{(0n)} \qquad\qquad (2.30)$$

Let $x \in R^n$ be an Euclidean point defined with $x = p_1 e_1 + p_2 e_2 + \cdots + p_n e_n$, then its hyperconformal form is given by Eq. 2.31.

$$X = \sum_{i=1}^{n} p_i e_i + \frac{1}{2}p_i^2 e_{(\infty i)} + e_{(0i)} \qquad\qquad (2.31)$$

In $G_{2n,n}$, the main geometric entity is the hyperellipse; we can define a hyperellipse with center in $(c_1, c_2, \cdots, c_n)$ and radii $r_1, r_2, \cdots, r_n$ as is shown in Eq. 2.32.

$$E = \sum_{i=1}^{n} \left[\frac{1}{r_i^2} (c_i e_i + e_{(0i)}) \right] + \frac{1}{2} \left[\sum_{i=1}^{n} \left(\frac{c_i^2}{r_i^2} \right) - 1 \right] e_\infty \qquad (2.32)$$

In the same way, all the other entities in the algebra shown in Table 2.2 can be derived as special cases of the hyperellipse, now taking into account that we have not only three radii (r_x, r_y, r_z), but until n, i.e. $r_i, i \in \{1, 2, \cdots, n\}$.

2.3 Clifford SVM

Clifford Support Vector Machines (CSVM) were presented in [17]; they are a generalization of the real-valued Support Vector Machines (SVM) [30] into multivector valued SVM. CSVM were designed to work with multi-vector-valued input data and to process this kind of input data taking advantage of the geometric features of the multivector spaces (geometric spaces) in which they are defined. CSVM can deal with multiple-class classification thanks to its Clifford product that represents the direct sum of linear spaces. So, we can treat each part of the input multivector in its subspace and then, using a multi-vector-valued kernel that uses the Clifford product, we can join the output in a single multivector to get a multi-output system. The CSVM solves a multivector valued primal optimization problem given by:

$$min \; L(w, b, \epsilon) = \frac{1}{2} a^T H a - \sum_{i,j} \alpha_{ij}, \qquad (2.33)$$

where the vector a is defined as:

$$
\begin{aligned}
a = \quad & [[a_{1_s}, a_{1_{e_1}}, a_{1_{e_2}}, \dots, a_{1_{e_1 e_2}}, \dots, a_{1_I}], \dots, \\
& , [a_{k_s}, a_{k_{e_1}}, a_{k_{e_2}}, \dots, a_{k_{e_1 e_2}}, \dots, a_{k_I}], \dots, \\
& , [a_{D_s}, a_{D_{e_1}}, a_{D_{e_2}}, \dots, a_{D_{e_1 e_2}}, \dots, a_{D_I}].
\end{aligned}
\qquad (2.34)
$$

are given by

$$
\begin{aligned}
a_{k_s}^T &= [(\alpha_{k_s})_1 (y_{k_s})_1, (\alpha_{k_s})_2 (y_{k_s})_2, ..., (\alpha_{k_s})_l (y_{k_s})_l], \\
a_{k_{e_1}}^T &= [(\alpha_{k_{e_1}})_1 (y_{k_{e_1}})_1, \dots, (\alpha_{k_{e_1}})_l (y_{k_{e_1}})_l], \\
& \qquad\qquad \cdot \\
& \qquad\qquad \cdot \\
& \qquad\qquad \cdot \\
a_{k_I}^T &= [(\alpha_{k_I})_1 (y_{k_I})_1, (\alpha_{k_I})_2 (y_{k_I})_2, ..., (\alpha_{k_I})_l (y_{k_I})_l];
\end{aligned}
\qquad (2.35)
$$

So, we can define the dual optimization problem as [2]:

$$max \qquad a^T 1 - \frac{1}{2}a^T Ha$$

$$subject\,to$$

$$0 \le (\alpha_s)_j \le C, \ 0 \le (\alpha_{e_1})_j \le C, ...,$$
$$0 \le (\alpha_{e_1 e_2})_j \le C, ..., 0 \le (\alpha_I)_j \le C$$
$$\text{for } j = 1, ..., l, \tag{2.36}$$

H is the **Gram** matrix, which is a positive semidefinite matrix that can be defined in terms of the matrices of the t-grade parts of $\langle x^{\dagger^T} x \rangle_t$ and is written as follows:

$$H = \begin{bmatrix} H_s H_{e_1} H_{e_2} .... H_{e_1 e_2} ... H_I \\ H_{e_1}^T H_s ... H_{e_4} .....H_{e_1 e_2} ... H_I H_s \\ H_{e_2}^T H_{e_1}^T H_s ... H_{e_1 e_2} ... H_I H_s H_{e_1} \\ . \\ . \\ . \\ H_I^T ... H_{e_1 e_2}^T ....H_{e_2}^T H_{e_1}^T H_s \end{bmatrix}. \tag{2.37}$$

The threshold $b \in \mathcal{G}_n$ can be computed by using the KKT conditions with the Clifford support vectors as follows:

$$b = b_s + b_{e_1}e_1 + ... + b_{e_1 e_2}e_1 e_2 + ... + b_I I$$

$$= \sum_{j=1}^{l}(y_j - w^{\dagger^T}x_j)/l. \tag{2.38}$$

subject to $a^T \cdot 1 = 0$; where a is given by (2.34) and, again, 1 denotes a vector of all ones.

For the nonlinear Clifford-valued classification problems, we require a Clifford algebra-valued kernel $K(x, y)$. In order to fulfill the Mercer theorem, we resort to a component-wise Clifford algebra-valued mapping

$$x \in G_n \xrightarrow{\phi} \Phi(x) = \Phi_s(x) + \Phi_{e_1}(x)e_1 + \Phi_{e_2}(x)e_2 + ...$$
$$+I\Phi_I(x) \in G_n.$$

In general, we build a Clifford kernel $K(x_m, x_j)$ by taking the Clifford product between the conjugated of x_m and x_j as follows:

$$K(x_m, x_j) = \Phi(x_m)^{\dagger}\Phi(x_j). \tag{2.39}$$

Next, as an illustration we present kernels using different geometric algebras.

---

[2]To get a detailed derivation of the dual problem, please see [17].

According to the Mercer theorem, there exists a mapping $\boldsymbol{u} : \mathcal{G} \to \mathcal{F}$, which maps the multivectors $\boldsymbol{x} \in \mathcal{G}_n$ into the complex Euclidean space:

$$\boldsymbol{x} \overset{v}{\to} u(\boldsymbol{x}) = u_r(\boldsymbol{x}) + I u_I(\boldsymbol{x}) \tag{2.40}$$

Recall that the center of a geometric algebra, i.e., $\{s, I = e_1 e_2\}$, is isomorphic with $\mathbb{C}$:

$$
\begin{aligned}
K(\boldsymbol{x}_m, \boldsymbol{x}_n) &= u(\boldsymbol{x}_m)^\dagger u(\boldsymbol{x}_n) \\
&= (u(\boldsymbol{x}_m)_s u(\boldsymbol{x}_n)_s + u(\boldsymbol{x}_m)_I u(\boldsymbol{x}_n)_I) + \\
&\quad + I(u(\boldsymbol{x}_m)_s u(\boldsymbol{x}_n)_I - u(\boldsymbol{x}_m)_I u(\boldsymbol{x}_n)_s), \\
&= (k(\boldsymbol{x}_m, \boldsymbol{x}_n)_{ss} + k(\boldsymbol{x}_m, \boldsymbol{x}_n)_{II}) + \\
&\quad + I(k(\boldsymbol{x}_m, \boldsymbol{x}_n)_{Is} - k(\boldsymbol{x}_m, \boldsymbol{x}_n)_{sI}) \\
&= \boldsymbol{H}_r + I \boldsymbol{H}_I.
\end{aligned}
\tag{2.41}
$$

For the quaternion-valued Gabor kernel function, we use $\boldsymbol{i} = e_2 e_3$, $\boldsymbol{j} = -e_3 e_1$, $\boldsymbol{k} = e_1 e_2$. The Gaussian window Gabor kernel function is

$$K(\boldsymbol{x}_m, \boldsymbol{x}_n) = g(\boldsymbol{x}_m, \boldsymbol{x}_n) exp^{-i \boldsymbol{w}_0^T (\boldsymbol{x}_m - \boldsymbol{x}_n)} \tag{2.42}$$

where the normalized Gaussian window function is given by

$$g(\boldsymbol{x}_m, \boldsymbol{x}_n) = \frac{1}{\sqrt{2\pi}\rho} exp^{-\frac{\|\boldsymbol{x}_m - \boldsymbol{x}_n\|^2}{2\rho^2}} \tag{2.43}$$

and the variables $\boldsymbol{w}_0$ and $\boldsymbol{x}_m - \boldsymbol{x}_n$ stand for the frequency and space domain respectively.

We can note that this kernel function keep separated the even and odd components of the given signal, i.e.,

$$
\begin{aligned}
K(\boldsymbol{x}_m, \boldsymbol{x}_n) &= K(\boldsymbol{x}_m, \boldsymbol{x}_n)_s + K(\boldsymbol{x}_m, \boldsymbol{x}_n)_{e_2 e_3} + \dots \\
&\quad + K(\boldsymbol{x}_m, \boldsymbol{x}_n)_{e_3 e_1} + K(\boldsymbol{x}_m, \boldsymbol{x}_n)_{e_1 e_2} \\
&= g(\boldsymbol{x}_m, \boldsymbol{x}_n) cos(\boldsymbol{w}_0^T \boldsymbol{x}_m) cos(\boldsymbol{w}_0^T \boldsymbol{x}_m) + \dots \\
&\quad + g(\boldsymbol{x}_m, \boldsymbol{x}_n) cos(\boldsymbol{w}_0^T \boldsymbol{x}_m) sin(\boldsymbol{w}_0^T \boldsymbol{x}_m) \boldsymbol{i} + \dots \\
&\quad + g(\boldsymbol{x}_m, \boldsymbol{x}_n) sin(\boldsymbol{w}_0^T \boldsymbol{x}_m) cos(\boldsymbol{w}_0^T \boldsymbol{x}_m) \boldsymbol{j} + \dots \\
&\quad + g(\boldsymbol{x}_m, \boldsymbol{x}_n) sin(\boldsymbol{w}_0^T \boldsymbol{x}_m) sin(\boldsymbol{w}_0^T \boldsymbol{x}_m) \boldsymbol{k}.
\end{aligned}
$$

Since $g(\boldsymbol{x}_m, \boldsymbol{x}_n)$ fulfills Mercer's condition, it is straightforward to prove that the $k(\boldsymbol{x}_m, \boldsymbol{x}_n)_u$ in the above equations satisfy these conditions as well.

After defining these kernels, we can proceed in the formulation of the SVM conditions. We substitute the mapped data $\Phi_i(\boldsymbol{x}) = \sum_{u=1}^{n} < \Phi_i(\boldsymbol{x}) >_u$ into the linear function $f(\boldsymbol{x}) = \boldsymbol{w}^{\dagger T} \boldsymbol{\Phi}(\boldsymbol{x}) + b$. The problem can be stated in a similar fashion to (2.33)-(2.38). In fact, we can replace the kernel function in

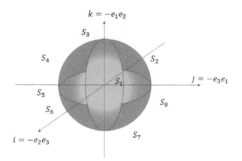

FIGURE 2.6: Quaternion valency states.

(2.36) to accomplish the Wolfe dual programming and thereby to obtain the kernel function group for nonlinear classification:

$$
\begin{aligned}
\boldsymbol{H_s} &= [K_s(\boldsymbol{x_m}, \boldsymbol{x_j})]_{m,j=1,\dots,l}\,, \\
\boldsymbol{H_{e_1}} &= [K_{e_1}(\boldsymbol{x_m}, \boldsymbol{x_j})]_{m,j=1,\dots,l}\,,
\end{aligned}
$$

$$
\vdots
$$

$$
\begin{aligned}
\boldsymbol{H_{e_n}} &= [K_{e_n}(\boldsymbol{x_m}, \boldsymbol{x_j})]_{m,j=1,\dots,l} \\
\boldsymbol{H_I} &= [K_I(\boldsymbol{x_m}, \boldsymbol{x_j})]_{m,j=1,\dots,l}\,.
\end{aligned}
\tag{2.44}
$$

Now, for the decision function used for classification, one uses the output function of the nonlinear Clifford SVM given by

$$
\boldsymbol{y} = csign_m\Big[f(\boldsymbol{x})\Big] = csign_m\Big[\boldsymbol{w}^{\dagger^T}\Phi(\boldsymbol{x}) + \boldsymbol{b}\Big],
\tag{2.45}
$$

where m stands for the state valency. That is, this is similar to employing the quadrature phase shift keying (QPSK) used in digital communication to encode different output values using two bits per symbol. So, we can use each part of the different grade of one multivector output to encode two (or more) values. For example, we can see in Fig. 2.6 the state of valency of the pure quaternion $csign_m$ function, where each state S_1 to S_8 correspond to a region of values for each part of the ith output multivector $y_i = y_{is} + y_{ii} + y_{ij} + y_{ik}$ for $\boldsymbol{i} = e_2e_3$, $\boldsymbol{j} = -e_3e_1$, $\boldsymbol{k} = e_1e_2 \in G_{0,3,0}$. So, Fig. 2.6 is the simile of one constellation diagram for QPSK that illustrates the quaternion valency states. We have to take into account that we can increase the number of classes that a single CSVM can classify because each part of the multivector output y_i can be an affecter for a $\pm$ sign. So, too, for the quaternion output case, se the CSVM can classify until $m \times 2^4$ classes

2.3.1 Quaternion valued support vector classifier

In [82] was presented an example of CSVM using the quaternion algebra of Hamilton $\boldsymbol{H}$, which is isomorphic with $G_{0,3,0}$, where $\boldsymbol{i} = e_2 e_3$, $\boldsymbol{j} = -e_3 e_1$, $\boldsymbol{k} = e_1 e_2$, named the Quaternion valued Support Vector Machine (QSVM). The polynomial quaternion-based kernel formulated as an extension of the real polynomial kernel was defined as

$$K\left(x_m, x_n\right) = \left(x_m^{\dagger}, x_n + c\right)^d \tag{2.46}$$

and also, it used the quaternion valued Gabor kernel function of Eq. 2.44. In the case of quaternions, the decision function for classification, Eq. 2.45, allows to classifying until $m \times 2^4 = 16$ classes solving one quadratic-valued optimization problem. The reader can review [82] for more technical detail. Also in [81] it is shown a parallel implementation of the CSVM using the Gaussian kernel that takes advantage of the features of this kernel to obtain a very efficient implementation of the CSVM approach is shown. We are going to use this implementation to obtain the experimental results shown in the next subsection.

2.3.2 Experimental results

In order to test the effectiveness of the CSVM to perform multiclassification tasks we are going to use a CSVM that works in $G_{0,3,0}$ algebra which is isomorphic to the quaternion algebra. Also, we use the methodology presented in [19] to use a the benchmark MNIST database of handwritten digits [70]. We use 60,000 handwritten digits (from zero to nine) for training and 10,000 for testing. We use a 7-fold cross validation methodology, and then we compute the mean error rate for the seven runnings of the classifiers.

We compare the results obtained using CSVM with a Gabor kernel Eq. 2.44 with a real SVM using the one-versus-one approach to do multiclassification since this real methodology was the one that showed better precision in the experiments of [17]. The real SVM employing the one-versus-one strategy was trained using the raw pixel values of the image of each digit, i.e. a vector of dimension $28 \times 28 = 784$. Meanwhile, for the CSVM we use, as in [19] the Fourier transform of each digit image and keep only the 100 most significant coefficients. As the coefficients of the Fourier transform are complex numbers we are only using the real s and the bivector part $e_2 e_3$ of the quaternion for the CSVM for each coefficient, i.e. $x_i = x_i(t)_s + y_i(t)e_2 e_3 = [x_i(t), y(i)_t, 0, 0]$.

The parameters used for the real SVM were fixed using a Gaussian kernel and a bio-inspired optimization algorithm named Particle Swarm Optimization (PSO) [10], using a swarm of size 40 particles. We obtain optimum parameters for the real SVM of C=1000 and $\sigma = .018$. Meanwhile, for the CSVM we use a C value of 1000.

Results obtained in the training stage using the 7-fold cross validation methodology were: a Mean Squared Error (MSE) of 2.98% for the real-SVM and 2.14% for the CSVM with the Gabor filter. Furthermore, we can compare the two methodologies using the total number of variables computed (TNV) per approach for doing multiclassification. The total number of variables computed is proportional to the number of quadratic problems that are solved (NQP) and the number of variables to compute per quadratic problem $NVQP$. So, using real-SVM with the one-versus-one approach, $NQP = K(K-1)/2 = 45$ where K is the number of classes that we want to recognize and $NVQP = 2*(N/K) = 2*(60,000/10) = 12,000$ where N is the total number of training examples (we have N/K training samples per class), to obtain a total number of variables to compute of $TNV = NQP * NVQP = 45 * 12,000 = 540,000$.

Meanwhile, using the CSVM, we have to solve one quadratic problem in total, i.e., $NQP = 1$ and the total number of variables (Lagrange multipliers) that we have to compute is $NVQP = D*N$; as we are working with complex input data, we have to compute a real s and an imaginary e_2e_3 Lagrange multiplier for each input vector, so $D = 2$ and $NVQP = 2*60,000 = 120,000$ So, the number of variables to compute using CSVM is considerably smaller than the TNV computed using real-SVM with the one-versus-one approach, this is directly reflected in the execution time, for the CSVM is about three times lower than for the real-SVM. With respect to the precision, the training MSE of the CSVM is even better than the one obtained with the real-SVM. The authors conclude that it is because we are considering all the training data when we solve the multivectorial-valued quadratic problem and we are working in the complex space where the input data is defined, so, we take the advantage that the geometry of the complex space is preserved.

Regarding the testing stage, the MSE obtained was 2.73% for the real-SVM and 2.01% for the CSVM for 10,000 data using the 7-fold cross-validation strategy.

In conclusion, the CSVM is an interesting approach to perform classification or object recognition for robots because it is a precise and fast algorithm that takes advantage when the input data can be described using geometric entities (nevertheless it is not necessary to use CSVM as classifiers). CSVM can be used as a multi-input, multi-output classifier system (MIMO) thanks to the multivectorial representation of the inputs and outputs. And it can solve one multivector valued optimization problem to perform multi-class classification instead of solving several quadratic optimization problems as it is doing using classical real-SVM approaches such as one-versus-one or one-versus-all.

Furthermore, the strategy presented in [19] can be used to preprocess any image and to obtain complex valued descriptors to be used as input data to CSVM which works solving the classification problem in the complex-valued vector space and takes advantage of this to obtain very accurate results.

2.4 Conformal Neuron and Hyper-Conformal Neuron

Geometric algebra has been used as a mathematical framework to design artificial neurons (and networks) with the objective of having more complex decision boundaries than the linear function defined when the perceptron (or Adaline) neuron is used. The first attempts to define geometric neurons were presented in [15, 21]. Nevertheless, the neurons presented in these approaches lacked explicit geometric interpretation. They take advantage of the algebra of Clifford more than the geometry (of the geometric algebras) when they use the Clifford product (geometric product for geometric algebras) in the propagation function instead of the inner product between vectors of weight and input, as is defined next

$$y = \sum_{i=1}^{n} w_i \otimes_{p,q,r} x_i + \theta \tag{2.47}$$

where w_i is the ith vector of synaptic weights, x_i is the ith vector of inputs, b is the bias, y_i is the ith output label and $\otimes$ is the geometric product for the geometric algebra $G_{p,q,r}$. The hypersphere neurons were presented in [11, 102] to deal with the issue of lacking explicit geometric interpretation, and to offer a neuron that, using the Conformal Geometric Algebra of Sec. 2.2.3 , computes a geometric boundary represented as a sphere (hypersphere for higher dimensions). The hypersphere neuron is a generalization of the perceptron because the planes (hyperplanes) computed by the perceptron are a special case of a deformed sphere when its radius is infinite.

2.4.1 Hyperellipsoidal neuron

The hyperellipsoidal neuron was presented in [133]. This neuron was designed using the GA $G_{6,3}$ of Sec.2.2.4 with the objective of having a neuron that computes quadratic surfaces as a class boundary without the need of using a kernel trick to achieve it. Although the Radial Basis Function (RBF) [124, 135] is capable of performing hyperellipsoidal classification, it uses the kernel trick of implicitly mapping input data into a higher dimensional feature space to be able to define quadratic surfaces as class boundaries. The hyperellipsoidal neuron also requires mapping the input data into a geometric algebra ($G_{6,3}$ for R^3 input data) but, in this space the activation function of the neuron is linear, and the computational complexity is reduced with respect to the RBF complexity. The map of input data that are in R^3 is defined in Eq. 2.25; meanwhile for a point in R^n the mapping is shown in Eq. 2.31, these $G_{2n,n}$ points are called hyperconformal points.

Since one hyperconformal point X and one hyperellipse E are 1-vectors in $G_{2n,n}^1$, their inner product is equivalent to the scalar product in the linear algebra of $R^{2n,n}$; therefore, taking into account the normalization in the null basis, we can find a better vectorial representation for the data point $x = (x_1, s_2, \cdots, x_n) \in R^n$, so, let us define the mapping $\psi_1 : R^n \to R^{2n+1}$ in equation 2.48

$$\psi_1(x) = X_c = \left[x_1, \cdots, x_n, 1, -\frac{1}{2}x_1^2, \cdots, -\frac{1}{2}x_n^2\right]^T \tag{2.48}$$

In the same way, for the hyperellipse E with parameters $(c_1, r_1, \cdots, c_n, r_n) \in R^{2n}$, we can find the mapping $\psi_2 : R^{2n} \to R^{2n+1}$ described in equation 2.49.

$$\psi_2(E) = E_c = \left[\frac{c_1}{r_1^2}, \cdots, \frac{c_n}{r_n^2},\right.$$
$$-\frac{1}{2}\left(\frac{c_1^2}{r_1^2} + \cdots + \frac{c_n^2}{r_n^2} - 1\right),$$
$$\left.\frac{1}{r_1^2}, \cdots, \frac{1}{r_n^2}\right]^T \tag{2.49}$$

Finally, the inner product between $\psi_1(x)$ in Eq. 2.48 and $\psi_2(E)$ in Eq. 2.49 can be rewritten as $X_c^T E_c$ where IP stands for Inner Product, this development is an extension of the development of the hypersphere in [11].

The scalar value resulting from the inner product computation between the hyperconformal point X and the hyperellipse E in IPNS represents the position of X in respect to E. When X lies on the surface of E the scalar value is equal to zero, and the sign of the scalar value represents that the point is inside or outside of the hyperellipse, so the decision function of the hyperellipsoidal neuron is

$$y = \text{sgn}(\psi_1(X)^T \psi_2(E)) \tag{2.50}$$

To train the hyperellipsoidal neuron we are going to use a hybrid methodology: the first step consists of using a clustering algorithm to train the center of the hyperellipse $\{c_1, c_2, \cdots, c_n\}$ as it is shown in [140]. Once the centers are fixed, we can adjust the radius using the next quadratic error definition

$$e = \frac{(y-d)^2}{2} \tag{2.51}$$

where y is the output of the neuron and d is the desired output.

Then we compute the gradient of the error in Eq. 6.21 with respect to the radius r_i

$$\frac{\delta e}{\delta r_i} = \frac{\delta e}{\delta y}\frac{\delta y}{\delta r_i} \tag{2.52}$$

and we perform an adjustment only if $y \neq 0$, so we compute y as

$$y = X_c^T E_c \tag{2.53}$$

$$= \frac{c_1 x_1}{r_1} + \frac{c_2 x_2}{r_2^2} + \cdots + \frac{c_n x_n}{r_n^2}$$

$$- \frac{1}{2} \left(\frac{c_1^2}{r_1^2} + \frac{c_2^2}{r_2^2} + \cdots + \frac{c_n^2}{r_n^2} - 1 \right)$$

$$- \frac{x_1^2}{2r_1^2} - \frac{x_2^2}{2r_2^2} - \cdots - \frac{x_n^2}{2r_n^2}$$

Taking into account just one radius at the time y is computed as

$$y_{r_i} = \frac{c_i x_i}{r_i^2} - \frac{c_i^2}{2r_i^2} - \frac{x_i^2}{2r_i^2} - 1 \tag{2.54}$$

$$= -\frac{(x_i - c_i)^2}{2r_i^2} - 1$$

Then, we can derive y_{r_i} with respect to one radius

$$\frac{\delta y}{\delta r_i} = \frac{\delta y_{r_i}}{\delta r_i} = \frac{(x_i - c_i)^2}{r_i^3} \tag{2.55}$$

and the final adjustment of the radius r_i is stated as

$$r_i \leftarrow r_i + \eta \frac{d}{r_i^3} (x_i - c_i)^2 \tag{2.56}$$

2.4.2 Experimental results

In this section, we are going to show the results of three classification problems in R^3 as input data dimension. In order to show the high capability of the hyperellipsoidal neuron (HN) to separate vectors of two classes, we design toy examples of classification problems and illustrate how the HN neuron adapts the classification boundary (decision surface) and deforms it into a 3D ellipse in Fig. 2.7, a cylinder in Fig. 2.8 or a pair of planes in Fig. 2.9 depending on the geometric position of the input data. It is important to note that in all the results shown, we obtained training error equal to zero.

2.5 Conclusions

In this chapter, we have shown two different approaches to design geometric machine learning algorithms that can be applied to perform object recognition

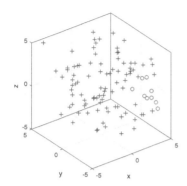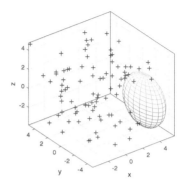

FIGURE 2.7: Left: input data, crosses and circles belong to different classes. Right: Classification result. The decision boundary was adapted as an ellipse.

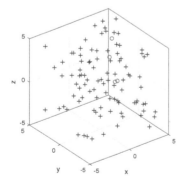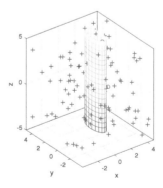

FIGURE 2.8: Left: input data, crosses and circles belong to different classes. Right: Classification result. The decision boundary was adapted as a cylinder.

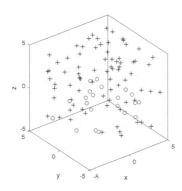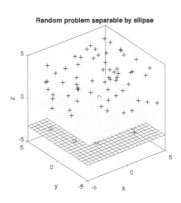

FIGURE 2.9: Left: input data, crosses and circles belong to different classes. Right: Classification result, the decision boundary was adapted as a pair of planes.

to aid robots to achieve a higher grade of autonomy when they are interacting with the environment. First we review the CSVM presented in [17] and show that they are an excellent choice to perform multi-class classification and that they can be applied to solve object recognition tasks to aid a robot to know how to interact with its environment. CSVMs are efficient classifiers that can be used as MIMO systems in order to do an efficient multi-class classification task because, no matter the number of classes to recognize, they solve one quadratic multivector-valued quadratic problem computing fewer variables (Lagrange multipliers) than classical approaches such as one-versus-one for real-SVMs. Therefore CSVM wastes less running time in training stages which is very useful to robotics that often have online execution restrictions. To solve just one quadratic problem to do multiclassification is possible because the Clifford product represents the direct sum of linear spaces and then we can treat each part of the input multivector in its subspace. Then, using a multivector-valued kernel that uses the Clifford product we can join the output in a single multivector to get a multi-output system.

Furthermore, the precision obtained using CSVM is high when we are processing complex and hypercomplex input data due to CSVMs taking advantage of the geometric properties of the multivector spaces in which this kind of data lie.

Second, we reviewed the conformal and hyperconformal neurons that were designed using conformal and hyperconformal geometric algebras with the objective of dealing with the issue of lacking explicit geometric interpretation that the classic artificial neurons have. Moreover, the hyperconformal neuron provides the ability to compute geometric class boundaries that can be adapted as a sphere ellipse, cylinder or pair of planes (hypersphere, hyperellipse, hypercylinder or pair of hyperplanes for higher dimensions), depending on the geometric distribution of the input data. The above is done without the need of using the kernel trick of machine learning algorithms such as RBF neural networks or SVM.

Also, it is important to note that the representation of objects through geometric entities can be a compact and shape-preserving description of them. So, representing objects with geometric entities using the mathematical framework of geometric algebra could be used to take advantage of geometric classifiers to improve its efficiency and effectiveness.

3

Non-Holonomic Mobile Robot Control Using Recurrent High Order Neural Networks

CONTENTS

3.1   Introduction

Most classical and modern control methodologies need to know the mathematical model of the system to be controlled. In some cases the systems are too complex to know their mathematical model with accuracy. In other cases, not all control techniques can be applied. The motivation to use a new control technique arises from the need to use mobile robots in disaster environments, in these environments the sensors of the robot are not always reliable, in addition to the robot being exposed to a large number of internal and external disturbances, the majority of which are not measurable. Therefore an inverse optimal neuronal controller is able to obtain the mathematical model implicitly and work under the presence of external disturbances and uncertainties and in turn adapt to the changes suffered by the robot [108].

In this work, we use as bases the works presented in [79, 108] where the recurrent high order neural network (RHONN) identifier and an inverse optimal control scheme for mobile robots are presented with just one simulation result. The RHONN is used to identify the plant model and it is trained with an extended Kalman filter EKF-based algorithm; under the assumption that the whole state is available for measurement. On the other hand, an inverse optimal control is designed for velocity trajectory tracking. So, in this work we extend [79] by including more simulation results, as well as real-time results using Maltlab[®][1] and two mobile robotics platforms in order to show the effectiveness of the proposed approach.

The neural identifier is based on the RHONN series-parallel model and it is used to get a mathematical model of a plant even with the presence of changing parameters; due to its on-line EKF-based training that adjusts the neural network weights during all the execution of the application [120]. This RHONN identifier has proven to be an effective identifier for real-time applications on systems where knowledge about the plant is unknown or insufficient or where their parameters change during operation [4, 6, 8, 105], besides, it presents good results even with time-delays [5].

On the other hand, the objective of optimal control is to determine the control signals which will force a process to satisfy physical constrains and at the same time minimize a performance criterion [66]. However, this requires solving the associated Hamilton Jacobi Bellman (HJB) equation which is not an easy task. This work uses the inverse approach in which solving the HJB equation is avoided. In this approach stabilizing feedback control is developed and then it is established that this control optimizes a cost functional [121].

The inverse optimal control is a Lyapunov function-based method as well as backstepping [56, 142, 143] and H∞ [27, 130]-based methods. These methods are adaptive and robust controllers and work well with uncertainties, however, in their design process they need a model of the system to be controlled. In the proposed RHONN identifier - inverse optimal control scheme, the identified model is used in the control process design. In this way, the main advantage of such a controller is that it does not require previous knowledge of the model of the system to be controlled.

Other advanced control methods for mechatronic systems to address the tracking trajectory and synchronization problems are vibration isolation for active suspensions with performance constraints and actuator saturation [131] and integrated adaptive robust control for multilateral teleoperation systems under arbitrary time delays [28]. However, there are not reported works for tracked robots; furthermore, in order to work, these methods need to know the model of the system to be controlled.

---

[1]MATLAB is a registered trademark of MathWorks Inc.

3.2 RHONN to Identify Uncertain Discrete-Time Non-linear Systems

Artificial neural networks, also known as neural networks, are massively parallel distributed processors built of a massive interconnection of simple computing units called neurons. They are designed to model the way in which the brain performs a task or a function [54], in other words, artificial neural networks are simplified models of the biological neural networks which we can implement in software or hardware [50, 54, 116]. Neural network computing power comes from its massive interconnection and its ability to learn and generalize. [54].

According to their architecture neural networks can be classified as in [50, 54, 108, 120]:

- *Static neural networks.* This kind of neural networks is capable of approximating any function using a static mapping.

- *Dynamic neural networks.* This type of neural networks has feedback connections which give them higher capability than static neural networks. Due to their feedback connection dynamic neural networks are capable of capturing the dynamic response of a system.

Neural networks are usually implemented using electronic components or are simulated in software [54].

The model of the RHONN used in this work is the one presented in Chapter 1, Section 1.2.

The Neural Network Training methodology was presented in Chapter 1, Section 1.2.2.3.

3.3 Neural Identification

Neural identification consists of selecting an appropriated neural network model and adjusting its weights according to an adaptation law, so that the neural network approximates the real system response for the same input [95].

Now, lets consider the following non-linear discrete time-delay MIMO system described by:

$$x(k + 1) = F(x(k - l), u(k))$$
$$y(k) = h(x(k)) \tag{3.1}$$

where $x \in \Re^n$, $u \in \Re^m$, $F \in \Re^n \times \Re^m \to \Re^n$ is a non-linear function and $l = 1, 2, \cdots$ is the unknown delay.

We use a model with time delay because the wireless communication with the tracked robots could induce some delays and we can modify the series-parallel model (1.1) to accept a state vector of a plan with time-delays [5], therefore, we get the following model:

$$\widehat{\chi}_i(k+1) = \omega_i^\top(k)z_i(x(k-l), u(k))$$
$$i = 1, 2, \cdots, n \tag{3.2}$$

This model is semi globally uniformly, ultimately, bounded and the proof can be found in [5].

The RHONN series-parallel (3.2) is selected to identify the model (3.1), and as adaptation law is used the extended Kalman filter-based algorithm (1.11) is used.

3.4   Inverse Optimal Neural Control

A desirable feature today is optimization; we want to work and use our time in an optimal way as we want our tools to work. The main objective of optimal control is to determine a control signal that allows that a process (the system) to satisfy some physical constraints and at the same time maximize (maximize or minimize) a performance criterion (cost function or performance index). Optimal control of nonlinear systems deals with obtaining a control law for a given system such that a cost functional is minimized [75, 121]. Dynamic programming, developed by Bellman is a solution for optimal control which leads to a nonlinear partial differential equation called the Hamilton-Jacobi-Bellman (HJB) equation. Solving this equation is not straightforward: for systems of dimension higher than two there are no practical ways to solve such an equation. Inverse optimal control is an alternative for the optimal control method which avoids the need to solve the associated HJB equation. For the inverse approach, a stabilizing feedback control law, based on a priori knowledge of a Control Lyapunov Function (CLF), is designed first and then it is established that this control law optimizes a cost functional. The mathematical model of a given system is required in order to implement the inverse optimal control law. In real-time applications, a controller based on the model of the system may not behave as desired because of model uncertainties; there are always internal and external disturbances, uncertain parameters and unmodelled dynamics. Neural networks have been established as an appropriate methodology for nonlinear function approximation; then, they can be employed for nonlinear system identification. Neural network adapts its synaptic weights in order to adjust its outputs to the system response [76]. Then to solve the trajectory tracking problem of uncertain discrete-time nonlinear systems, in this chapter we deal first with the inverse optimal control, and therefore this controller is applied to a RHONN model in order to obtain an Inverse Optimal Neural Controller (IONC).

Consider the following affine discrete nonlinear system (3.3)

$$\mathbf{x}(k+1) = \mathbf{f}(\mathbf{x}(k)) + \mathbf{g}(\mathbf{x}(k))\mathbf{u}(k) \tag{3.3}$$

where $\mathbf{x} \in \Re^n$ is the state of the system, $\mathbf{u} \in \Re^m$ the control input, $\mathbf{f} : \Re^n \to \Re^n$ and $\mathbf{g} : \Re^n \to \Re^{n \times m}$ are smooth maps. System (3.3) is supposed to have an equilibrium point $\mathbf{x}(0) = 0$. Moreover, the full state $\mathbf{x}(k)$ is assumed to be available.

For the inverse optimal control a Lyapunov control function is designed to satisfy the passivity condition which states that a passive system can be stabilized by making a negative feedback from the output. $u(k) = -\alpha y(k)$ with $\alpha > 0$.

Equation 3.4 is proposed as a control Lyapunov fuction [75, 121] in order to ensure stability of the system (3.3).

$$V(\mathbf{x}(k)) = \frac{1}{2}\mathbf{x}(k)^\top P\mathbf{x}(k), \quad P = P^\top > 0 \tag{3.4}$$

Instead of solving the associated HJB equation, the inverse optimal control synthesis is based on the knowledge of $V(\mathbf{x}(k))$. The inverse optimal control law for system (3.3) with (3.4) is (3.5).

$$
\begin{aligned}
u(k) &- -\frac{1}{2}R^{-1}(\mathbf{x}(k))g^\top(\mathbf{x}(k))\frac{\partial V(\mathbf{x}(k))}{\partial \mathbf{x}(k+1)} \\
&= -\frac{1}{2}(R(\mathbf{x}(k)) + \frac{1}{2}g^\top(\mathbf{x}(k))Pg(\mathbf{x}(k)))^{-1} \times \\
&\quad g^\top(\mathbf{x}(k))P\mathbf{f}(\mathbf{x}(k))
\end{aligned} \tag{3.5}
$$

where $R(\mathbf{x}(k)) = R(\mathbf{x}(k))^\top > 0$ is a matrix whose elements can be functions of the system state or can be fixed. P is a matrix such that the inequality (3.6) holds.

$$
\begin{aligned}
V_f(\mathbf{x}(k)) &- \frac{1}{4}P_1^\top(\mathbf{x}(k))(RP(\mathbf{x}(k)))^{-1}P_1(\mathbf{x}(k)) \leq \\
&- \mathbf{x}^\top(k)Q\mathbf{x}(k)
\end{aligned} \tag{3.6}
$$

with

$$RP(\mathbf{x}(k)) = R(\mathbf{x}(k)) + P_2(\mathbf{x}(k)) \tag{3.7}$$

$$V_f(\mathbf{x}(k)) = \frac{1}{2}\mathbf{f}^\top(\mathbf{x}(k))P\mathbf{f}(\mathbf{x}(k)) - V(\mathbf{x}(k)) \tag{3.8}$$

$$P_1(\mathbf{x}(k)) = g^\top(\mathbf{x}(k))P\mathbf{f}(\mathbf{x}(k)) \tag{3.9}$$

$$P_2(\mathbf{x}(k)) = \frac{1}{2}g^\top(\mathbf{x}(k))Pg(\mathbf{x}(k)) \tag{3.10}$$

$$Q = Q^\top > 0 \tag{3.11}$$

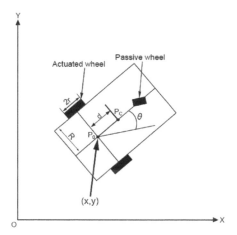

FIGURE 3.1: Schematic model of a mobile robot with two actuated wheels, where x, y are the coordinates of P_0, θ is the heading angle of the mobile robot.

In [121], it is demonstrated that the control law (3.5) is globally asymptotically stable. Moreover, (3.5) is inverse optimal in the sense that it minimizes a cost functional [121].

3.5   IONC for Non-Holonomic Mobile Robots

In this section the IONC is applied to two kinds of mobile robots: wheeled and tracked robots; both of them have non-holonomic restrictions for movement, complicating the trajectory tracking problem. Besides, in this work an electrically actuated model is included for the robot model.

3.5.1   Robot model

We consider a mobile robot with two actuated wheels as shown in Fig. 3.1, as follows:

The kinematics of an electrically driven differential robot is described by the state-space model (3.12) [31, 88, 108].

$$\dot{x}_1 = J(x_1)x_2 \tag{3.12}$$
$$\dot{x}_2 = M^{-1}(-C(\dot{x}_1)x_2 - Dx_2 - \tau_d + NK_Tx_3)$$
$$\dot{x}_3 = L_a^{-1}(u - R_ax_3 - NK_Ex_2)$$

each subsystem is defined by (3.13):

$$x_1 = [x_{11}, x_{12}, x_{13}]^\top = [\mathbf{x}, \mathbf{y}, \theta]^\top$$

$$x_2 = [x_{21}, x_{22}]^\top = [\mathbf{v}_2, \mathbf{v}_1]^\top \tag{3.13}$$

$$x_3 = [x_{31}, x_{32}]^\top = [\mathbf{i}_{\mathbf{a}_1}, \mathbf{i}_{\mathbf{a}_2}]^\top$$

$$u = [\mathbf{u}_1, \mathbf{u}_2]^\top$$

where $\mathbf{x}$ and $\mathbf{y}$ are the coordinates of P_0, θ is the heading angle of the robot (Fig. 3.1), $\mathbf{v}_1$ and $\mathbf{v}_2$ are the angular velocities of the robot, $\mathbf{i}_{\mathbf{a}_1}$ and $\mathbf{i}_{\mathbf{a}_2}$ are the currents of the motor of the robot, $\mathbf{u}_1$ and $\mathbf{u}_2$ are input voltages, x_3 is the actuator dynamics.

where

$$J(x_1) = 0.5r \begin{bmatrix} \cos(x_{13}) & \cos(x_{13}) \\ \sin(x_{13}) & \sin(x_{13}) \\ R^{-1} & -R^{-1} \end{bmatrix} \tag{3.14}$$

$$M = \begin{bmatrix} x_{11} & x_{12} \\ x_{12} & x_{11} \end{bmatrix} \tag{3.15}$$

$$N = \begin{bmatrix} n_1 & 0 \\ 0 & n_2 \end{bmatrix} \tag{3.16}$$

$$K_T = \begin{bmatrix} K_{t_1} & 0 \\ 0 & K_{t_2} \end{bmatrix} \tag{3.17}$$

$$L_a = \begin{bmatrix} l_{a_1} & 0 \\ 0 & l_{a_2} \end{bmatrix} \tag{3.18}$$

$$R_a = \begin{bmatrix} r_{a_1} & 0 \\ 0 & r_{a_2} \end{bmatrix} \tag{3.19}$$

$$K_E = \begin{bmatrix} K_{e_1} & 0 \\ 0 & K_{e_2} \end{bmatrix} \tag{3.20}$$

where R is half of the width of the tracked robot and r is the radius of the wheels which drive the tracks. M is the inertia matrix symmetric and positive defined by the physical parameters of the tracked robot, d is the distance from the center of mass P_c of the mobile robot to the middle point P_0 between the right and left driving wheels. m_c and m_w are the masses of the body and wheel with a motor, respectively. I_c, I_w and I_m are the moment of inertia of the body about the vertical axis through P_c, the wheel with a motor about the wheel axis, and the wheel with a motor about the wheel diameter, respectively. The positive terms d_{ii}, $i = 1, 2$, are the damping coeficients, $\tau_d \in R^2$ is a vector of disturbances including unmodeled dynamics, which is considered bounded. K_T is the motor torque constant, L_a is the inductance, K_E is the back electromotive force coefficient and R_a the resistance of the actuator. Here, disturbances are considered unknown but bounded so that $|\tau_{d_i}| \leq d_{mi}$, $i = 1, 2$.

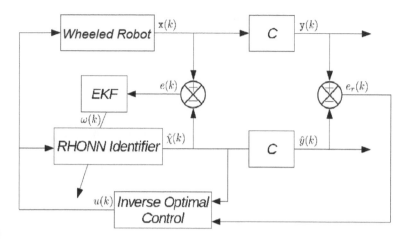

FIGURE 3.2: Identifier-control scheme for wheeled differential robots.

3.5.2   Wheeled robot

The main problem for trajectory tracking of non-holonomic mobile robots is that they possess mobility restrictions; therefore, it is necessary to design error surfaces to guarantee a correct tracking of the trajectories generated by computational algorithms. The variation of parameters and the inaccurate models that are developed for this kinds of system are other problems to consider; the resistance in the rotors of the actuators increases due to the temperature. There are also other parameters that change over time and therefore the modeling error may be greater. Since the development of models in general tends to be scary and often does not agree with reality, controllers have recently been developed without the need of a model of the plant. It is only necessary to know the relative degree or system order. The concern has arisen about applying these kind of generic controllers to mobile robots with the advantage that the control is designed only once and can be applied to different mobile robots of the same type. As the implementation of the controllers typically is performed using digital devices such as processors, programmable logic devices and microcontrollers, these controllers should be designed in discrete-time to optimize and guarantee stability in such scenarios where a sampling period must be included.

Then in this section, robust controllers are proposed to solve the trajectory tracking problem of electrically driven nonholonomic wheeled mobile robots.

3.5.2.1   Controller design

The kinematics of an electrically driven tracked robot is described by the state-space model, described in Eq. 3.12.

Figure 3.2 shows the closed-loop for the neural identifier-inverse optimal control scheme.

3.5.2.2 Neural identification of a wheeled robot

The discrete-time RHONN identifier (3.21) is proposed to get a valid mathematical model for tracked mobile robots.

$$
\begin{aligned}
\hat{\chi}_1(k+1) &= \omega_{11}(k)S(x_{11}(k)) + \omega_{12}(k)S(x_{12}(k)) \\
&\quad + \omega_{13}(k)S(x_{13}(k)) + \omega_{14}(k)x_{21}(k) \\
&\quad + \omega_{15}(k)x_{22}(k) \\
\hat{\chi}_2(k+1) &= \omega_{21}(k)S(x_{11}(k)) + \omega_{22}(k)S(x_{12}(k)) \\
&\quad + \omega_{23}(k)S(x_{13}(k)) + \omega_{24}(k)x_{21}(k) \\
&\quad + \omega_{25}(k)x_{22}(k) \\
\hat{\chi}_3(k+1) &= \omega_{31}(k)S(x_{11}(k)) + \omega_{32}(k)S(x_{12}(k)) \\
&\quad |\ \omega_{33}(k)S(x_{13}(k)) + \omega_{34}(k)x_{21}(k) \\
&\quad + \omega_{35}(k)x_{22}(k) \\
\hat{\chi}_4(k+1) &= \omega_{41}(k)S(x_{11}(k)) + \omega_{42}(k)S(x_{12}(k)) \\
&\quad + \omega_{43}(k)S(x_{21}(k)) + \omega_{44}(k)S(x_{31}(k)) \\
&\quad + \omega_{45}(k)x_{31}(k) \\
\hat{\chi}_5(k+1) &= \omega_{51}(k)S(x_{11}(k)) + \omega_{52}(k)S(x_{12}(k)) \\
&\quad + \omega_{53}(k)S(x_{22}(k)) + \omega_{54}(k)S(x_{32}(k)) \\
&\quad + \omega_{55}(k)x_{32}(k) \\
\hat{\chi}_6(k+1) &= \omega_{61}(k)S(x_{11}(k)) + \omega_{62}(k)S(x_{12}(k)) \\
&\quad + \omega_{63}(k)S(x_{21}(k)) + \omega_{64}(k)S(x_{31}(k)) \\
&\quad + \omega_{65}(k)u_1(k) \\
\hat{\chi}_7(k+1) &= \omega_{71}(k)S(x_{11}(k)) + \omega_{72}(k)S(x_{12}(k)) \\
&\quad + \omega_{73}(k)S(x_{22}(k)) + \omega_{74}(k)S(x_{32}(k)) \\
&\quad + \omega_{75}(k)u_2(k)
\end{aligned}
\tag{3.21}
$$

where $\hat{\chi}_1$, $\hat{\chi}_2$, $\hat{\chi}_3$, $\hat{\chi}_4$, $\hat{\chi}_5$, $\hat{\chi}_6$ and $\hat{\chi}_7$ identify x, y, θ, v_1, v_2, i_{a_1} and i_{a_2}, respectively. Moreover, it is worth mention that this model includes the actuator dynamics.

The RHONN identifier (3.21) is adapted on-line using the EKF-based training algorithm (1.11). All the neural network states and weights are initialized in a random way.

3.5.2.3 Inverse optimal control of a wheeled robot

The control objective is the design of a control law u to track the desired trajectory generated by the following reference robot:

$$
\begin{aligned}
\dot{x}_r &= v_r \cos(\theta_r) \\
\dot{y}_r &= v_r \sin(\theta_r) \\
\dot{\theta}_r &= \omega_r
\end{aligned}
$$

where x_r, y_r and θ_r are the position and orientation of the reference robot. v_r and ω_r are the linear and angular velocity of the reference robot, respectively.

The design of a controller based on the model (3.12) requires the exact knowledge of the plant parameters and disturbances which can vary with time.Since in practice getting this model is not a trivial task, the identified model (3.21) is used as a valid model for the tracked robots; therefore, the identified model (3.21) is used to design the controller for solving the trajectory tracking problem .

Systems (3.12) and (3.22) are discretized based on the Euler methodology. System (3.12) is rewritten in the block structure form (3.22) in order to simplify the controller synthesis.

$$
\begin{aligned}
\hat{\chi}_a(k+1) &= \omega_1(k)z_a(k) + \omega_a(k)x_b(k) \\
\hat{\chi}_b(k+1) &= \omega_2(k)z_b(k) + \omega_b(k)x_c(k) \\
\hat{\chi}_c(k+1) &= \omega_3(k)z_c(k) + \omega_c(k)u_2(k)
\end{aligned}
\tag{3.22}
$$

where

$$
\begin{aligned}
\hat{\chi}_a &= [\hat{\chi}_1, \hat{\chi}_2, \hat{\chi}_3]^\top \\
\hat{\chi}_b &= [\hat{\chi}_4, \hat{\chi}_5]^\top \\
\hat{\chi}_c &= [\hat{\chi}_6, \hat{\chi}_7]^\top
\end{aligned}
\tag{3.23}
$$

$$
\begin{aligned}
x_b &= x_2 \\
x_c &= x_3
\end{aligned}
\tag{3.24}
$$

with

$$
\omega_1(k) = \begin{bmatrix} \omega_{11} & \omega_{12} & \omega_{13} \\ \omega_{21} & \omega_{22} & \omega_{23} \\ \omega_{31} & \omega_{32} & \omega_{33} \end{bmatrix}
\tag{3.25}
$$

$$
\omega_a(k) = \begin{bmatrix} \omega_{14} & \omega_{15} \\ \omega_{24} & \omega_{25} \\ \omega_{34} & \omega_{35} \end{bmatrix}
\tag{3.26}
$$

$$
\omega_2(k) = \begin{bmatrix} \omega_{41} & \omega_{42} & \omega_{43} & \omega_{44} \\ \omega_{51} & \omega_{52} & \omega_{53} & \omega_{54} \end{bmatrix}
\tag{3.27}
$$

$$
\omega_b(k) = \begin{bmatrix} \omega_{45} & 0 \\ 0 & \omega_{55} \end{bmatrix}
\tag{3.28}
$$

$$\omega_3(k) = \begin{bmatrix} \omega_{61} & \omega_{62} & \omega_{63} & \omega_{64} \\ \omega_{71} & \omega_{72} & \omega_{73} & \omega_{74} \end{bmatrix} \tag{3.29}$$

$$\omega_c(k) = \begin{bmatrix} \omega_{65} & 0 \\ 0 & \omega_{75} \end{bmatrix} \tag{3.30}$$

In this way, the control objective is to force x_a to track the reference signal $x_{a\delta}(k+1) = [x_r, y_r, \theta_r]$ which is achieved by designing a control law $x_b(k) = u_1(k)$ based on an inverse optimal approach for discrete-time nonlinear affine systems [121]. Also, x_b is forced to track the previous control law which is achieved designing a control law as (3.31):

$$x_c = \omega_b^{-1}(-\omega_2(k)z_b(k) + u_1(k)) \tag{3.31}$$

Therefore, the reference signal for the control law u_2 is $x_{c\delta}(k+1) = x_c$. Thus, the inverse optimal control laws are defined as:

$$u_i(k) = -[I_m + J_i(x(k))]^{-1} h_i(x(k), x_\delta(k+1)) \tag{3.32}$$

with

$$h_i(x_i(k), x_{i\delta}(k+1)) = g_i^\top(x(k)) P_i(f_i(x_i(k)) \tag{3.33}$$
$$- x_{i\delta}(k+1))$$

$$J_i(x(k)) = \frac{1}{2} g_i^\top(x(k)) P_i g_i(x(k))$$

where $i = 1, 2$.

3.5.2.4 Experimental results

In order to show the applicability of the proposed control scheme, a real-time implementation of the control strategy for a mobile robot is presented. This implementation is performed in a *QBot*® robot, which is shown in Fig. 3.3. This experiment is performed with a sampling time of 0.06s. The results are presented as follows: Fig. 3.4 shows the trajectory tracking performance for the mobile robot, Fig. 3.5 presents input signal to each wheel, Fig. 3.6 shows the trajectory tracking and identification results for x, y and θ, respectively, and finally the identification error is included in Fig. 3.7. It is important to note that the plant parameters are unknown.

3.5.3 Tracked robot

A tracked robot is a mobile robot that runs on continuous tracks instead of wheels. The main advantage of a tracked robot is that it can be used to navigate in rough terrains [45, 139].

The thrust developed by a wheeled vehicle will generally be lower than that

FIGURE 3.3: Wheeled differential robot.

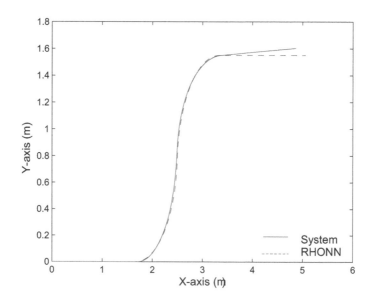

FIGURE 3.4: Trajectory tracking performance for the mobile robot.

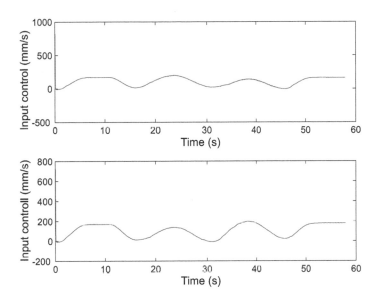

FIGURE 3.5: Applied control signals.

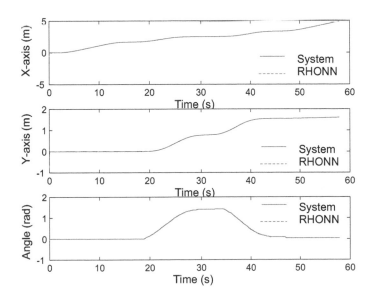

FIGURE 3.6: Trajectory tracking and identification results for x, y and θ.

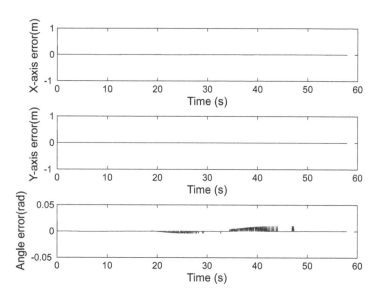

FIGURE 3.7: Identification error for x, y and θ.

developed by a comparable tracked vehicle [139], which is why these kinds of vehicles are used in a variety of applications where terrain conditions are difficult or unpredictable: urban reconnaissance, forestry, mining, agriculture, rescue mission scenarios, autonomous planetary explorations, to name but a few [45, 126]. Besides, tracked robots offer some other advantages, such as:

- Tracked robots are versatile vehicles in different terrains and weather conditions.

- Tracked robots generate low ground pressure which conserves the environment.

- The design of tracked robots prevents them from sinking, or becoming stuck in soft ground.

Tracked mobile robots can be considered as the most important type of mobile robots and an extensive class of controller has been proposed for trajectory tracking in this kind of robots [1, 41, 88, 136]. These works have the common characteristic that they need to know a lot of information about the system to be controlled, and most of them are implemented in continuous time. It is well known that the current trend towards digital rather than analog control of dynamic systems is mainly due to the advantages found in working with digital rather than continuous-time signals [96].

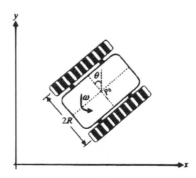

FIGURE 3.8: Schematic model of a tracked mobile robot, where x, y are the coordinates of P_0, θ is the heading angle of the mobile robot.

3.5.3.1 Controller design

Figure 3.8 shows a tracked robot configuration:

The kinematics of an electrically driven tracked robot is described by the state-space model, described in Eq. 3.12. Figure 3.9 shows the closed-loop for the neural identifier-inverse optimal control scheme. The design of the identifier-controller is the same for the tracked robot as the design for the wheeled robot in Section 3.5.2.

3.5.3.2 Results

This section presents the simulation and real-time results. The simulations of the model (3.12), the RHONN identifier (3.21) and the control (3.32) were implemented in MATLAB and Simulink[2] software. On the other hand, in the real-time tests, the block that represents the model of the robot was removed and replaced with a block where the communication with the HD2® was implemented.

In this way, it was demonstrated that the same RHONN identifier is capable of identifying both models, the one used for simulation and the one for the HD2. Also, due to the fact that the control was designed using the model of the RHONN identifier, it is valid for both simulation and real-time tests.

The parameters for the RHONN identifier for all tests are shown in Table 3.1.

Moreover, the following weights are fixed: $\omega_{45} = 1$, $\omega_{55} = 1$, $\omega_{65} = 0.00001$ and $\omega_{75} = 0.00001$. The initial posture for the reference robot is $x_r = 0$, $y_r = 0$ and $\theta_r = 0$.

---

[2]Simulink is a registered trademark of MathWorks Inc.

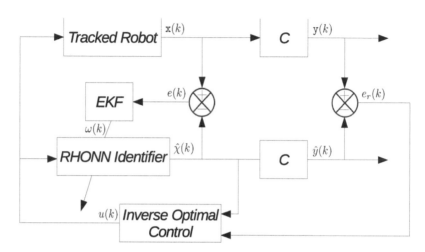

FIGURE 3.9: Control scheme.

TABLE 3.1: EKF-based training algorithm parameters.

i	$\mathbf{P}$	$\mathbf{Q}$	$\mathbf{R}$
1	$diag(3 \times 10^6)$	$diag(3 \times 10^6)$	1×10^4
2	$diag(3 \times 10^6)$	$diag(3 \times 10^6)$	1×10^4
3	$diag(3 \times 10^6)$	$diag(3 \times 10^6)$	1×10^4
4	$diag(4 \times 10^4)$	$diag(4 \times 10^6)$	1×10^3
5	$diag(4 \times 10^4)$	$diag(4 \times 10^6)$	1×10^3
6	$diag(4 \times 10^4)$	$diag(4 \times 10^6)$	1×10^3
7	$diag(4 \times 10^4)$	$diag(4 \times 10^6)$	1×10^3

TABLE 3.2: RMSE of simulation test tracking errors of x, y and θ.

	Real signals		
	x	y	θ
RMSE	0.0239	0.0189	0.0128
	Identified Signals		
	x	y	θ
RMSE	0.0241	0.0211	0.0134

Simulation - Test 1.

The parameters for the test are set as follows:

$$P_1(k) = 14400 \begin{bmatrix} 162 & 1 & 2 \\ 1 & 162 & 3 \\ 2 & 3 & 162 \end{bmatrix}$$

$$P_2(k) = 20 \begin{bmatrix} 1 & 0 \\ 0 & 1 \end{bmatrix}$$

$$g_1 = 0.5rT \begin{bmatrix} \cos(x_{13}) & \cos(x_{13}) \\ \sin(x_{13}) & \sin(x_{13}) \\ R^{-1} & -R^{-1} \end{bmatrix}$$

$$g_2 = \begin{bmatrix} 1 & 0 \\ 0 & 1 \end{bmatrix}$$

$$T = 0.001s \tag{3.34}$$

where T is the sample time.

Figures 3.10 and 3.11 show the references for linear and angular velocities.

Figures 3.12, 3.13 and 3.14 show the references, real and identified signals of x, y and θ.

Errors of the simulation test are shown in Figure 3.15 and Table 3.2. Figure 3.15 shows the errors of reference vs simulated real signals, a reference vs identified real signals is omitted because the error between the real and identified signal is so small that a second figure showing the reference vs identified signal would look just like Figure 3.15.

Figures 3.16 and 3.17 show the velocities v_1 and v_2, respectively.

Figures 3.18 and 3.19 show the currents i_1 and i_2, respectively.

Figures 3.20 and 3.21 show the control signals.

Table 3.3 shows the root mean square (RMSE) of the real *vs* identified signals.

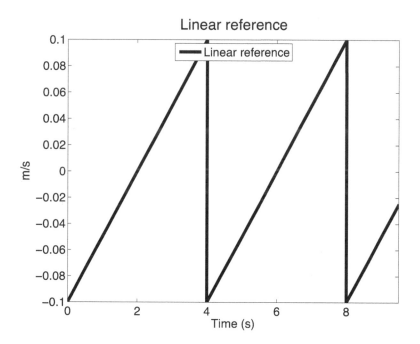

FIGURE 3.10: Linear velocity reference.

TABLE 3.3: RMSE of simulated real *vs* identified state variables of simulation Test 1.

i	**RMSE**
1	0.0031
2	0.0049
3	0.0029
4	0.0088
5	0.0068
6	0.0767
7	0.0128

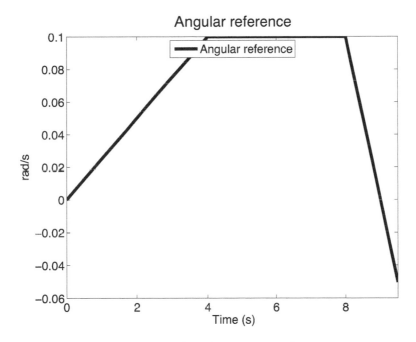

FIGURE 3.11: Angular velocity reference.

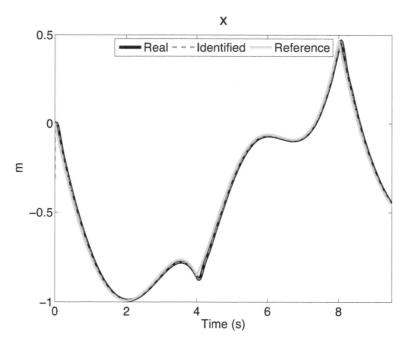

FIGURE 3.12: Real, identified and reference signals of x.

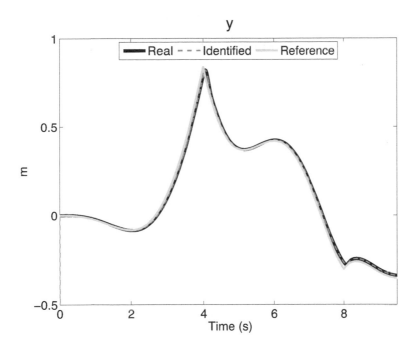

FIGURE 3.13: Real, identified and reference signals of y.

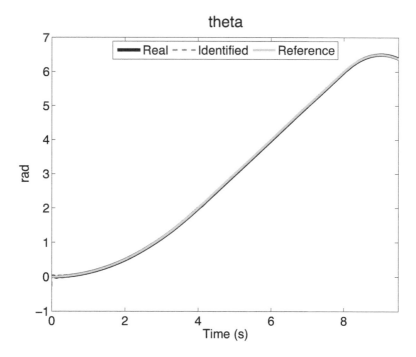

FIGURE 3.14: Real, identified and reference signals of θ.

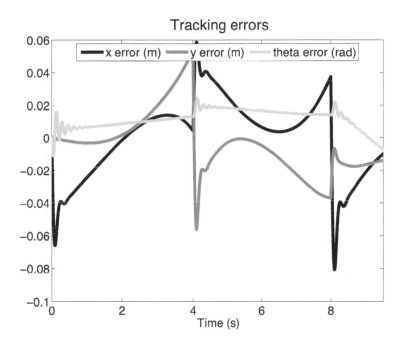

FIGURE 3.15: Tracking errors of simulation test.

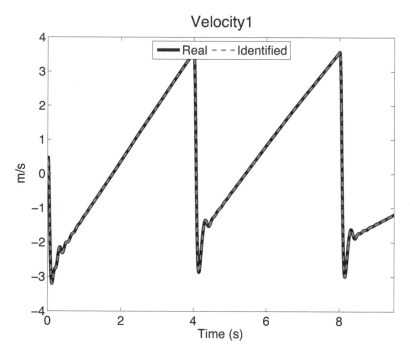

FIGURE 3.16: Real and identified signals of v_2.

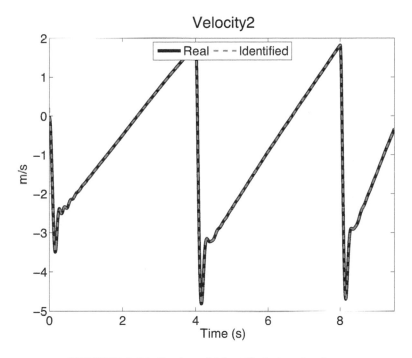

FIGURE 3.17: Real and identified signals of v_2.

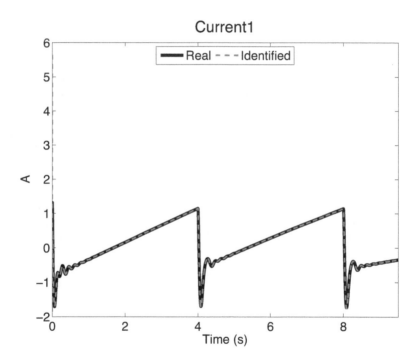

FIGURE 3.18: Real and identified signals of i_1.

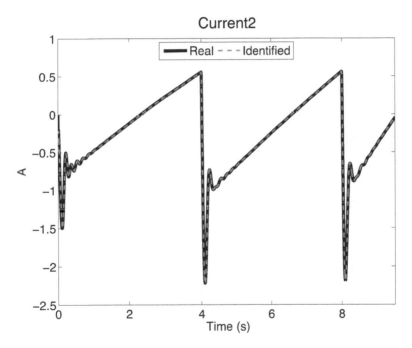

FIGURE 3.19: Real and identified signals of i_2.

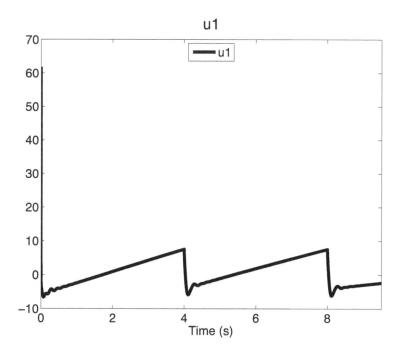

FIGURE 3.20: Control signal u_1.

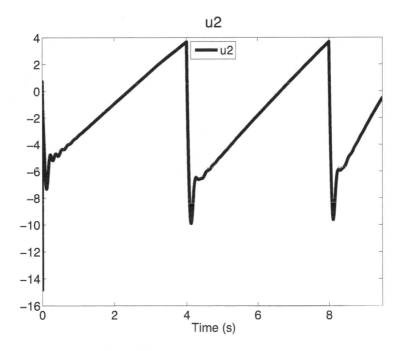

FIGURE 3.21: Control signal u_2.

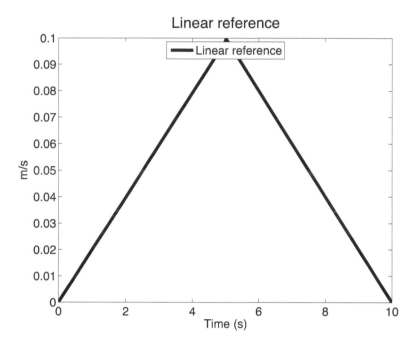

FIGURE 3.22: Linear velocity reference.

Performance comparison

The following simulation results compare the proposed neural identifier - inverse optimal control scheme and the super twisting control proposed in [77] which is a discrete-time control algorithm for nonholonomic wheeled mobile robots, without the previous knowledge of the plant model or its parameters. The super twisting control proposed in [77] was adapted from the three states model presented in [77] to our seven states model.

For the comparison we use the same references shown in Figures 3.22 and 3.23.

Tracking performance for the neural identifier - inverse optimal control is shown in Figures 3.24, 3.25 and 3.26, for x, y and θ, respectively. Tracking performance for super twisting is shown in Figures 3.27, 3.28 and 3.29, for x, y and θ, respectively.

For more details, Figures 3.30 and 3.31 show the errors and Table 3.4 shows the RMSEs of the tests.

For the previous Figures 3.22 to 3.31 and Table 3.4, a better performance is found in the neural identifier - inverse optimal control scheme than the super twisting in states x, and especially in θ where an undesirable behavior is presented when it is used the super twisting controller is used.

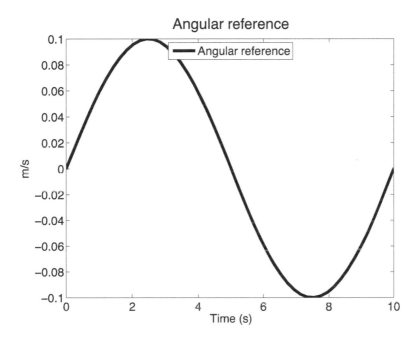

FIGURE 3.23: Angular velocity reference.

TABLE 3.4: RMSE of simulation comparison test tracking errors of x, y and θ.

		Real signals - Inverse optimal control		
		x	y	θ
RMSE		0.0225	0.0348	0.0126
		Identified signals - Inverse optimal control		
		x	y	θ
RMSE		0.0260	0.0362	0.0158
		Real signals - Super Twisting		
		x	y	θ
RMSE		0.0317	0.0036	0.0652

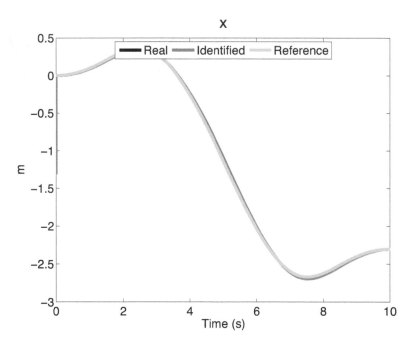

FIGURE 3.24: Real, identified and reference signals of x with neural identifier - inverse optimal control.

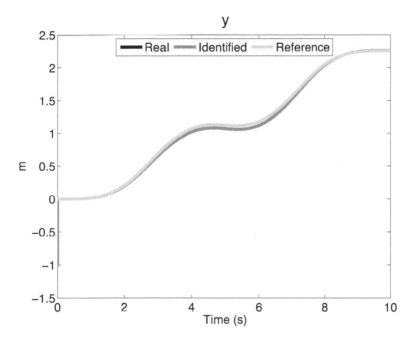

FIGURE 3.25: Real, identified and reference signals of y with neural identifier - inverse optimal control.

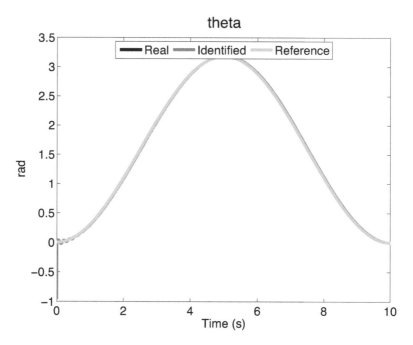

FIGURE 3.26: Real, identified and reference signals of θ with super twisting.

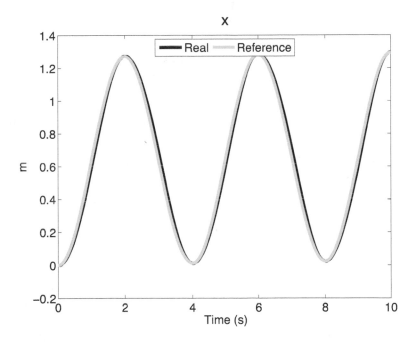

FIGURE 3.27: Real and reference signals of x with super twisting.

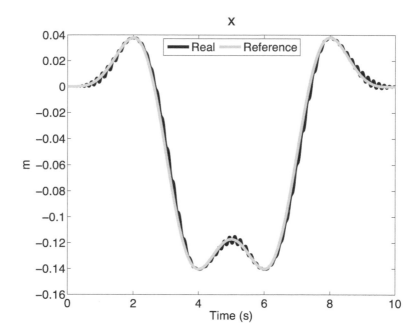

FIGURE 3.28: Real and reference signals of y with super twisting.

Real-Time Results

Figure 3.32 shows the HD2 Treaded ATR Tank Robot Platform with an added wireless router.

Figure 3.33 shows the HD2 Treaded ATR Tank Robot Platform inside where the modification of the platform can be seen. Basically, the modification is the replacement of the original board with a systems base on Arduino[3] and added current sensors; Figure 3.33 also shows the original batteries and motors.

Real Time - Test 1.

The parameters for the real time test are set as follows:

$$P_1(k) = 72000 \begin{bmatrix} 162 & 1 & 2 \\ 1 & 162 & 3 \\ 2 & 3 & 162 \end{bmatrix}$$

$$P_2(k) = 10000 \begin{bmatrix} 1 & 0 \\ 0 & 1 \end{bmatrix}$$

---

[3] Arduino is a registered trademark of Arduino LLC.

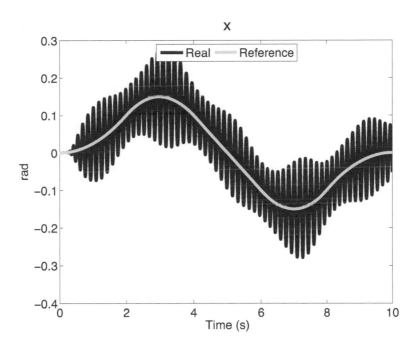

FIGURE 3.29: Real and reference signals of *theta* with super twisting.

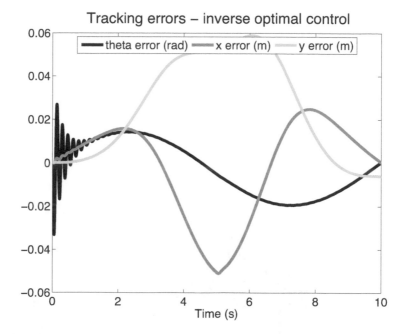

FIGURE 3.30: Tracking errors with neural identifier - inverse optimal control.

$$g_1 = 0.5rT \begin{bmatrix} \cos(x_{13}) & \cos(x_{13}) \\ \sin(x_{13}) & \sin(x_{13}) \\ R^{-1} & -R^{-1} \end{bmatrix}$$

$$g_2 = \begin{bmatrix} 1 & 0 \\ 0 & 1 \end{bmatrix}$$

$$T = 0.003s \tag{3.35}$$

Figures 3.34 and 3.35 show the references for linear and angular velocities.

Figures 3.36, 3.37 and 3.38 show the references, real and identified signals of x, y and θ.

In contrast to simulation results the presented real-time tracking results of Figures 3.36 - 3.38 cannot achieve a perfect tracking mainly to the following reason: wifi communication and packet loss, actuators saturation, noise and precision of the sensors, unmodeled dynamics

However, they have the same dynamic and small errors shown in Figure 3.39 and Table 3.5. Also, it is important to mention that this is accomplished without knowledge of our modified tracked robot model and parameters.

Figures 3.40 and 3.41 show the velocities v_1 and v_2, respectively.

Figures 3.42 and 3.43 show the currents i_1 and i_2, respectively.

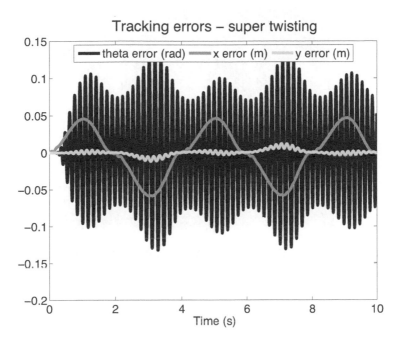

FIGURE 3.31: Tracking errors with super twisting.

FIGURE 3.32: HD2 Treaded ATR Tank Robot Platform.

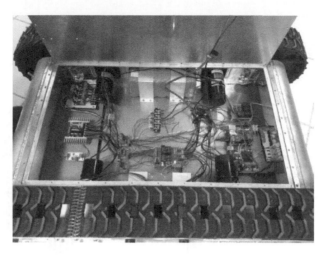

FIGURE 3.33: HD2 Treaded ATR Tank Robot inner components.

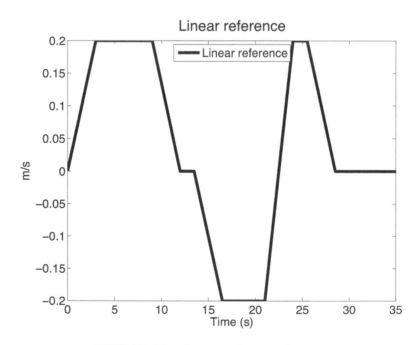

FIGURE 3.34: Linear velocity reference.

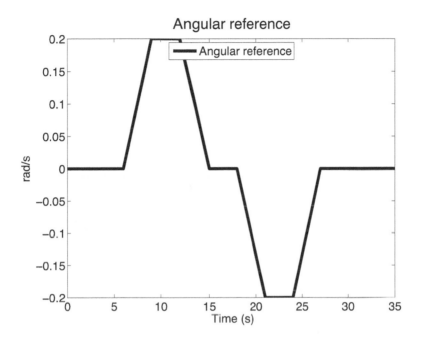

FIGURE 3.35: Angular velocity reference.

TABLE 3.5: RMSE of real-time Test 1 tracking errors of x, y and θ.

	Real signals		
	x	y	θ
RMSE	0.0216	0.0098	0.0182
	Identified Signals		
	x	y	θ
RMSE	0.0238	0.0101	0.0193

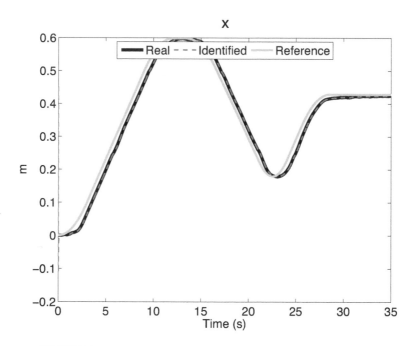

FIGURE 3.36: Real, identified and reference signals of x.

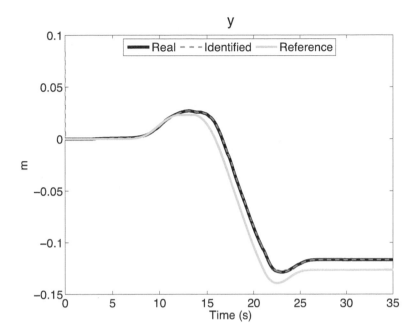

FIGURE 3.37: Real, identified and reference signals of y.

Figures 3.44 and 3.45 show the control signals.

Table 3.6 shows the root mean square (RMSE) of the real vs identified signals.

Real Time - Test 2.

The parameters for this test are the same as the previous test. Figures 3.46 and 3.47 show the references for linear and angular velocities.

Figures 3.48, 3.49 and 3.50 show the references, real and identified signals of x, y and θ.

Errors of real-time Test 2 are shown in Figure 3.51 and Table 3.7.

Figures 3.52 and 3.53 show the velocities v_1 and v_2, respectively.

Figures 3.54 and 3.55 show the currents i_1 and i_2, respectively.

Figures 3.56 and 3.57 show the control signals.

Table 3.8 shows the root mean square (RMSE) of the real vs identified signals.

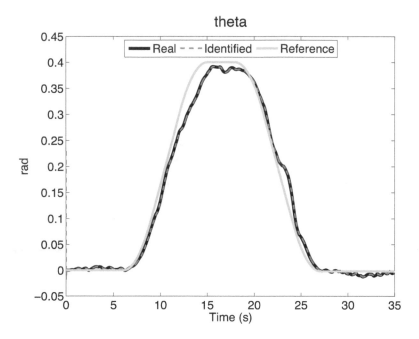

FIGURE 3.38: Real, identified and reference signals of θ.

TABLE 3.6: RMSE of real *vs* identified states of real-time Test 1.

i	RMSE
1	0.0071
2	0.0013
3	0.0035
4	0.0280
5	0.0204
6	0.1389
7	0.1309

TABLE 3.7: RMSE of real-time Test 2 tracking errors of x, y and θ.

	Real signals		
	x	y	θ
RMSE	0.0378	0.0101	0.0309
	Identified Signals		
	x	y	θ
RMSE	0.0388	0.0179	0.0342

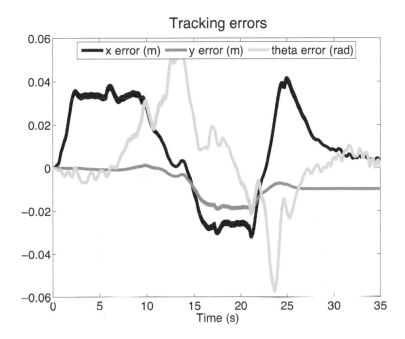

FIGURE 3.39: Tracking errors of real-time Test 1.

TABLE 3.8: RMSE of real vs identified states of real-time Test 2.

i	RMSE
1	0.0026
2	0.0146
3	0.0098
4	0.0343
5	0.0210
6	0.1510
7	0.1404

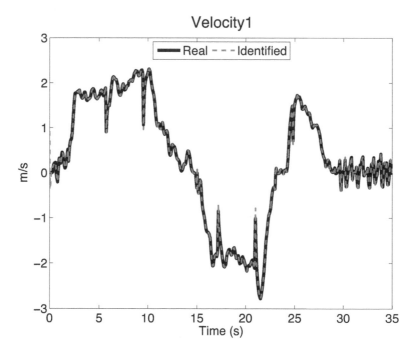

FIGURE 3.40: Real and identified signals of v_1.

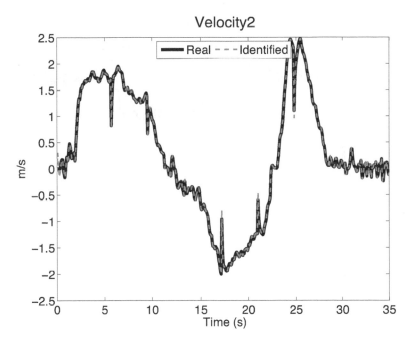

FIGURE 3.41: Real and identified signals of v_2.

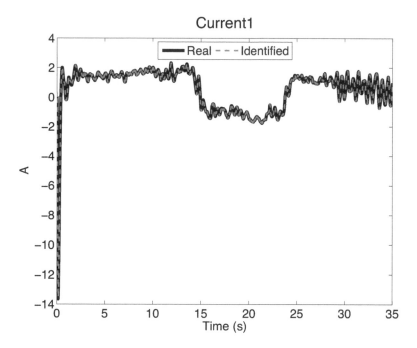

FIGURE 3.42: Real and identified signals of i_1.

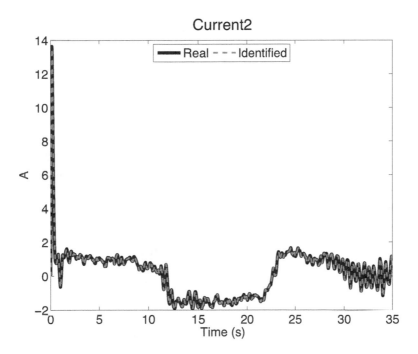

FIGURE 3.43: Real and identified signals of i_2.

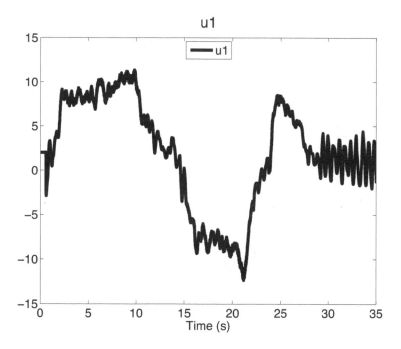

FIGURE 3.44: Control signal u_1.

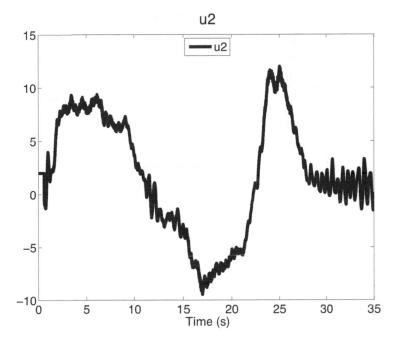

FIGURE 3.45: Control signal u_2.

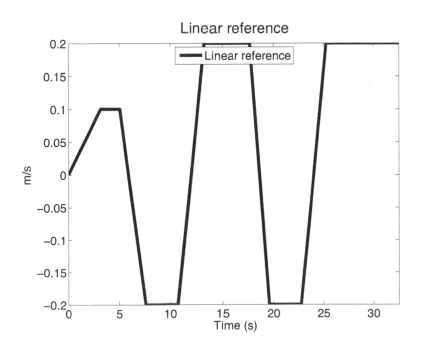

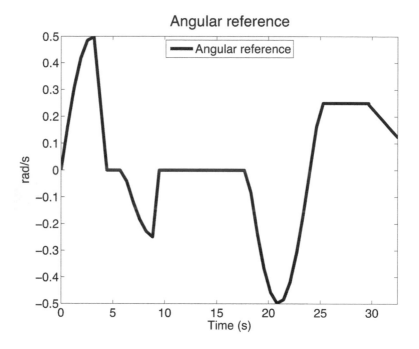

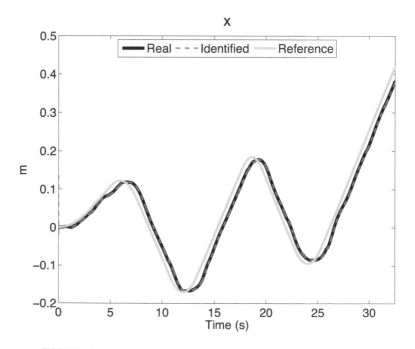

FIGURE 3.46: Real, identified and reference signals of x.

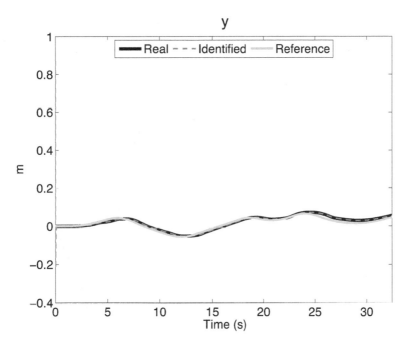

FIGURE 3.47: Real, identified and reference signals of y.

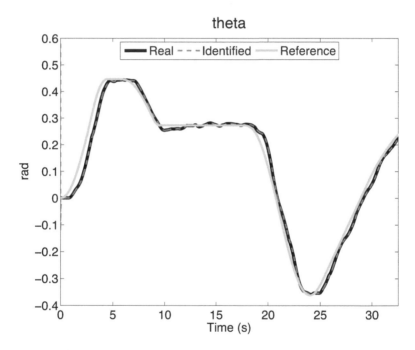

FIGURE 3.48: Real, identified and reference signals of θ.

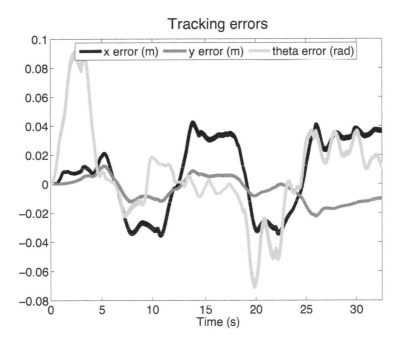

FIGURE 3.49: Tracking errors of real-time Test 2.

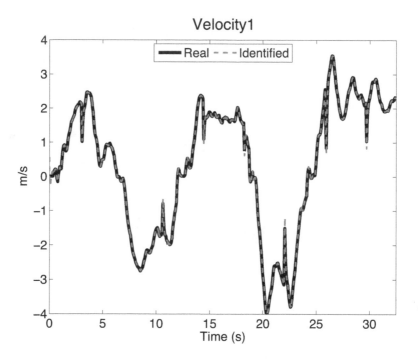

FIGURE 3.50: Real and identified signals of v_1.

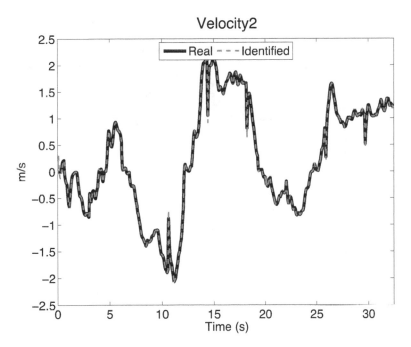

FIGURE 3.51: Real and identified signals of v_2.

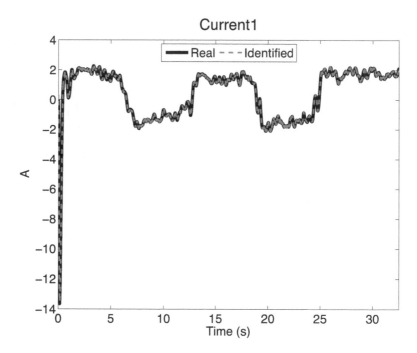

FIGURE 3.52: Real and identified signals of i_1.

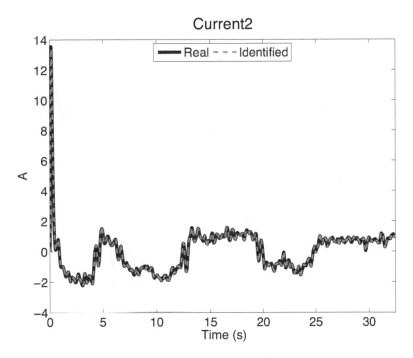

FIGURE 3.53: Real and identified signals of i_2.

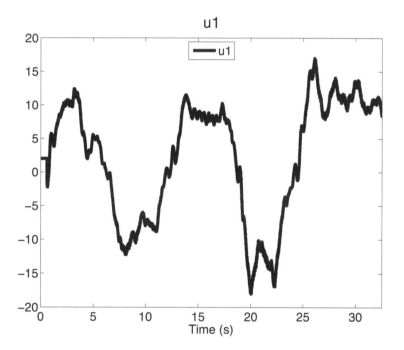

FIGURE 3.54: Control signal u_1.

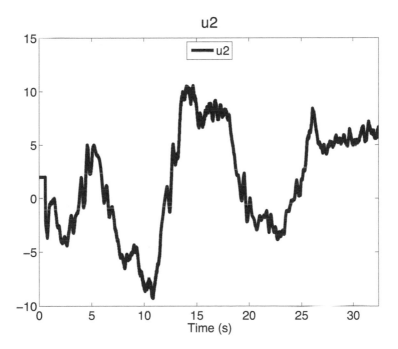

FIGURE 3.55: Control signal u_2.

3.6   Conclusions

This work uses a RHONN identifier trained with a EKF-based algorithm to get the model of mobile robots, first for a wheeled differential robot and then for an all terrain tracked robot, for real-time in both cases. Through the results, it can be seen that the same neural identifier is capable of identifying both simulated and real models. It is important to say that knowledge about the model parameters of our mobile robots is considered unknown for the controller design. Moreover, for the simulation even if the parameter were available they were not used for the design of the identifier on real-time due to the parametric time-variations. On the other hand, the inverse optimal control which has been designed using the identified model shows good results for simulation in real time tests. This can be appreciated in the result graphs which compare the references against the real signals of both mobile robots used.

4

Neural Networks for Autonomous Navigation on Non-Holonomic Mobile Robots

CONTENTS

4.1   Introduction

For autonomous movement, a robotic system must be able to interact with the environment and recognize, reconstruct, take and execute an appropriate action using a low level control system to accomplish its goal [33, 35, 36]. However, for a robot to be truly independent and able to cope with real environments, it has to solve many sub-tasks [43].

This work shows the integration of two approaches: the implementation of robot navigation systems and an efficient method of control composed by a neural identifier for the problem of path planning [94]; this arises based on the need for mobile robots on wrecked environments preferably avoiding human intervention either because it is highly repetitive or dangerous [3].

The system implements Simultaneous Localization and Mapping (SLAM) during the exploration phase to make a reconstruction of the environment, when the system faces a dynamic environment in which it is hard to locate itself through SLAM, it uses experience previously obtained with Reinforcement Learning (RL) to guide the robot to a goal with the correct action (created by a indirect controller formed by a neural identifier and an inverse optimal controller). The new environment information which is obtained by navigating with RL is added to the map constructed with the SLAM algorithm [125]. This allows for an autonomous navigation system to face widely changing environments over time [127]. Then the integration of planning (RL-SLAM)-identifier-controller is implemented and applied in real time using a differential robot. The vehicle is comprised of an iRobot Create® robotic platform and an array of infrared and sonar sensors, with wireless communication.

4.2   Simultaneous Localization and Mapping

In this section is presented a subsystem dedicated to solving the tasks of path planning and mapping.

The problem that the SLAM algorithms solve is about exploring an unknown environment, while simultaneously a map of this environment is being created using sensor information and the navigating agent is localizing itself on this map. A lot of techniques have been proposed to solve this problem so far. The first steps towards a full solution of the SLAM problem were proposed between 1985 and 1986 in the works of Smith and Durrant-Whyte. They established the geometric uncertainty and its effect on the environmental feature (landmark) location. At that time, the efforts to include methods based on statistics and estimation theory in robotics and Artificial Intelligence began. Nicholas Ayache and Oliver Faugeras, Raja Chatila and Jean-Paul Lamond designed the first solutions for the SLAM problem based on the Kalman filter for visual robot navigation. The 1986 IEEE International Conference on Robotics and Automation (ICRA) allowed a great discussion about the different mapping techniques thanks to the works of Smith, Self and Cheeseman [9, 78].

This discussion was the basis for the formal approach to the SLAM problem because it proved that the localization error caused a generalized error on the feature (landmark) perception of the environment. This is to say, that all the location error were correlated. This statement established the need to correct not only the position of the robot but the entire map at the same time. This would require a very large state vector which would include all the estimated positions of the robot, and this would represent a very heavy computational memory cost. Also, at that time there was no proof of the convergence of this approach, and then it was assumed that the estimation error of the map would grow unlimited. This caused the SLAM research to be delayed until 1995 when M. Csorba [35] showed that the SLAM problem would converge when it was turned into a complete problem, that is to say, when it was solved the problem of localization and mapping at the same time and not as separated tasks. In 1999 in the International Symposium on

Robotics Research (ISRR) the first discussion on SLAM was held, and some proofs of convergence between SLAM methods based on the Kalman Filter and probabilistic mapping were shown. Then, in the year 2000, at the ICRA Conference a lot of attention was paid to the SLAM theme, and the main issues in obtaining a complete solution to the SLAM problem were very clear, such as data association, computationally efficient implementations, implementations on multiple robots as well as implementations for outdoor, indoor, air and submarine environments [9, 78].

The features recognized by SLAM algorithms are considered by definition, static over time. The SLAM problem requires a probability distribution function which is calculated for all time k. This probability distribution describes the posterior density of the location of both the robot and the characteristics map at a time k [125, 137].

$$P(x_{(k)}, m | Z_{(0:k)}, U_{(0:k)}, x_{(0)}) = \frac{P(z_{(k)} | x_{(k)}, m) P(x_{(k)}, m | Z_{(0:k-1)}, U_{(0:k)}, x_{(0)})}{P\left(z_{(k)} | Z_{(0:k)}, U_{(0:k)}\right)}$$

(4.1)

where $x_{(k)}$ is the robot position at the time k, m is the map of the environment, $Z_{(0:k-1)}$ and $U_{(0:k-1)}$ are the sequence of observations and the sequence of control actions respectively, from time 0 to time $k - 1$, $x_{(0)}$ is the initial position of the robot, this is known as the Bayes filter, and is a popular form of inference mapping for its recursive formulation, allowing additional observations to be incorporated into the posterior efficiently.

EKF-SLAM is the first highly accepted SLAM implementation; it is based on the Extended Kalman Filter because it offers features such as, the representation of the problem in the form of a state-space model with additive Gaussian noise, and the employment of linearized models of nonlinear motion and observation models.

SLAM implementations that use the Kalman filter must take the same characteristics as the filter requires, like: the kinematic model of the robot, the environment model, and data association mode. Once you have these basic requirements, the implementation of EKF-SLAM consists of three basic stages: prediction, observation and status update [35, 36].

4.2.1 Prediction

The prediction is carried out by passing the last estimate of the pose at time k based on the control signal $u_{(k)}$ and using the information generated prior to time k.

$$x_{(k+1|k)} = f(x_{(k|k+1)}, u_{(k)}) + w_{(k)}$$

(4.2)

where f models vehicle kinematics and $w_{(k)}$ models mean white noise, then the joint distribution over the map and robot poses is $P(x_{(k)}, m | Z_{(0:k-1)}, U_{(0:k-1)},$

$x_{(0)}$). Furthermore, if many features m_i in the map m are correlated by observations taken from a single pose, then inferring the true posterior may become computationally demanding.

4.2.2   Observations

Each time k, measurements are obtained with a sensor that generates the vector $Z_{(k)}$. In the algorithm, the map m consists of a set of features $m_1, m_2, ..., m_n$, such as grid cells or landmark position. The map is inferred from a sequence of observations $Z_{(0:k)} = z_{(0)}, z_{(1)}, ..., z_{(k)}$, where $z_{(k)}$ is the observation at time k; these observations are generally measurements of the features m_i, and the subscript $0 : k$ denotes a sequence of measurements taken at discrete intervals from time $= 0$ to some end time $= k$. Each observation $z_{(k)}$ is modeled as a stochastic function $P(z_{(k)}|x_{(k)}, m)$ of both the map and the vehicle pose at time k, denoted as $x_{(k)} = (x_{(k)}, y_{(k)}, \theta_{(k)})$ and it is named as the observation function or measurement model. The observation model is described as

$$P(z_{(k)}|x_{(k)}, m) \leftrightarrow z_{(k)} = h(x_{(k)}|m) + v_{(k)} \tag{4.3}$$

where h describes the geometry of the observation and where $v_{(k)}$ are additive uncorrelated Gaussian observation errors.

4.2.3   Status update

Estimation is performed using a gain matrix which provides a weighted sum of the prediction and observation.

$$P_{(k)} = P_{(k|k-1)} - W_{(k)} S_{(k)} W_{(k)}{}^t \tag{4.4}$$

$$W_{(k)} = P_{(k|k-1)} \left(\frac{\partial h}{\partial x} \right)_{(k)} S_{(k)}{}^{-1} \tag{4.5}$$

$$S_{(k)} = \left(\frac{\partial h}{\partial x} \right)_{(k)} P_{(k|k-1)} \left(\frac{\partial h^t}{\partial x} \right)_{(k)} + R_{(k)} \tag{4.6}$$

where P is the covariance matrix of the robot pose (landmarks-landmarks, landmarks-robot position), R is the covariance observation error, and the Jacobian of h ($\frac{\partial h}{\partial x}$) evaluated at $x_{(k-1)}$. Once the SLAM system determines that a new feature has been observed, it is added to the map in the state vector increasing its size. It is equally necessary to increase the covariance matrix.

4.3 Reinforcement Learning

Reinforcement Learning is a strategy based on the interaction of a system or agent with its environment, allowing it to learn to perform a task automatically. RL defines a relation between situations and actions to maximize a numerical reward generated by the response of the environment. RL begins with a complete system that involves the environment and a definite goal [132]. The task usually is a series of actions that the robot has to perform to achieve its goal, then, the mission of the learner is to find the action rules (policies) to optimally achieve a certain goal through its interaction with the environment. What distinguishes RL from other learning methods is that information used to train the system is obtained through evaluation of the results obtained by the actions taken. This requires an active exploration and a trial and error search approach to find the best performance that determines how good is an action taken or what is the best course of action in a given situation [127, 132].

The key elements for the RL [61, 132] are:

1. The environment.

2. The approach; rules/policies that define the actions to take for guiding the robot to the desired goal.

3. The reward; it defines the goal to achieve, relating each state-action that the system can take.

4. The cost function; it determines the best actions to achieve the final task defining the value of a state s with the highest amount of reward that the system can obtain from that state to the final state.

This method consists of representing the environment where the agent is conducting its task as a discrete space consisting of state-action pairs $(s_{(k)}, a_{(k)})$. The dynamics of the system are: From a state $s_{(k)}$ the action $a_{(k)}$ selected from the set A is performed by an agent, as a result, the agent receives a reward with an expected value $R(s_{(k)}, a_{(k)})$ and the current state changes to the following state $s_{(k+1)}$ according to the probability transition $P(s_{(k+1)}|s_{(k)}, a_{(k)})$. The agent continues selecting and executing actions, creating a path of states visited until it arrives at the desired position.

$$R = R(s_{(0)}, a_{(0)}) + \gamma R(s_{(1)}, a_{(1)}) + R(s_{(2)}, a_{(2)}) + ... \tag{4.7}$$

$$R = \sum_{n=0}^{\infty} \gamma^n r_{(k+n+1)} \tag{4.8}$$

γ is the forgetting factor used to weight the importance the system gives to long term rewards against immediate rewards. Later, the sum of expected rewards obtained by the agent in tests is maximized.

$$V(s) = E(R_{(k)}|s_{(k)}) \qquad (4.9)$$

Then, the cost function can be defined generally as

$$V(s) = E\left(R_{(k)}|s_{(k)}\right) = E(\sum_{n=0}^{\infty} \gamma^n r_{(k+n+1)}) \qquad (4.10)$$

The used RL algorithm is the one known as Q-Learning. The optimal Q value is defined as the sum of rewards obtained by performing an action on a state and following the optimal policy as $V(s)$[61].

$$Q(s_{(k)}, a_{(k)}) = R(s_{(k)}, a_{(k)}) + \gamma max(Q(s_{(k+1)}, a_{(k)})) \qquad (4.11)$$

Only the action with the highest Q-value is taken. It contains information about the number of states between the current pose and the goal, and the traversability-cost of each state. The Q-Learning algorithm is implemented to obtain an intelligent exploration agent with capabilities to learn and to deal with dynamic environments while it is mapping and locating itself in its environments.

With the obtained information, the system is capable of getting a path and constructing a map of the environment; however, to move the agent mobile robot a control system is needed. Such system must give the correct action as it moves through the path designed by the RL.

Q-Learning can be expressed in the algorithm (1).

Algorithm 1 Q-Learning

For each (s, a), initialize $Q(s, a)$ to zero.
Observe the current state $s_{(k)}$
while current state $\neq$ Goal **do**
    Select an action $a_{(k)}$ from state $s_{(k)}$ and execute it.
    Receive immediate reward $r_{(k+1)}$.
    Observe the new state $s_{(k+1)}$.
    Update the entry for $Q(s_{(k)}, a_{(k)})$.
    Update state $s_{(k)} = s_{(k+1)}$.
end while

4.4   Inverse Optimal Neural Controller

To deal with the problem of following the path planned in the map created in the above section, the low control system is implemented using a neural network approach that is presented in this section. When neural network control approaches are presented, it is generally understood that a neural network is responsible for calculating the control action, but it can be divided into two groups: direct control and indirect control. In the first method, the control is performed by the neural network; on the other hand, indirect control is always based on models and the objective is to use a neural network to identify the system model [134]. The plant information is obtained by running the application in order to acquire a lot of data to describe the system behavior. This process consists in obtaining the parameters that best make the association between inputs and outputs. The goal of this stage is focused on giving the system a known input and observing how the system output behaves [134].

As stated in Chapter 3, the main purpose of an optimal control is to obtain a control signal that causes the process to satisfy some physical restrictions [67, 93].

For the inverse optimal controller the Lyapunov Control Function (LCF) (4.12) is designed in order to satisfy the passivity condition, which states that a passive system can be stabilized by making a negative feedback from the output $u_{(k)} = \alpha y_{(k)}$, con $\alpha > 0$ [98].

$$V(x_{(k)}, x_{ref(k)}) = \frac{1}{2}(x_{(k)} - x_{ref(k)})^T K^T P K (x_{(k)} - x_{ref(k)}) \qquad (4.12)$$

where $x_{ref(k)}$ is the desired path and K is a gain matrix further introduced to modify the rate of convergence of the tracking error [20, 98].

This solution is applied on the neural identifier developed in Chapter 3 to obtain a discrete-time neural model for an electrically driven nonholonomic mobile robot (Chapter 3) in Eq. (3.21) with control law (3.31).

4.4.1   Planning-Identifier-Controller

Classic SLAM deals with environments which are considered by definition, static over time; however in dynamic environments, a SLAM algorithm must somehow manage moving objects. So that it can detect and ignore them; it may track them as moving landmarks, but it must not add a moving object to the map and assume it is stationary. The conventional SLAM solution is highly redundant [7]. Simultaneous estimation of moving and stationary landmarks is very costly due to the added predictive model. For this reason, the implemented solution presented in this paper involves a stationary EKF SLAM update combined with a simple RL module to deal with moving objects.

RL-SLAM integrates SLAM algorithm and RL. For the integration of the systems, we use RL as a control system ; this is the origin of the actions that direct the robot and will be the entries $u_{(k)}$ to the SLAM estimation. Measurements $z_{(k)}$ are used in addition to the SLAM estimated location of the robot for the creation of a discrete space-state to be employed by RL to obtain the path to follow. This relationship is explained as follows: the RL navigation system could not perform its task without the map drawn by the SLAM, and SLAM itself could not generate this map without the control actions generated by the RL control system. Furthermore, SLAM itself can not interpret maps; SLAM can not allow the agent learning policies to help it to avoid high cost states-action pairs.

Algorithm 2 Planning-Identifier-Controller

1: Establish initial state and goal.
2: Obtain initial desired route Q.
3: Initialize SLAM map $P_{(0)} = 0$, initial pose.
4: Get observations $z_{(0)}$.
5: Add new features.
6: **while** current state $(x_{(k)}) \neq$ Goal **do**
7:     Update state map with $z_{(k)}$, $x_{(k)}$.
8:     Update $Q_{(k)}$ based on the observed.
9:     Search obstacles.
10:     **if** obstacle **then**
11:         Look for a new route in $Q_{(k)}$ and update map.
12:         **if** No path found in $Q_{(k)}$ **then**
13:             It is not possible to find a route.
14:         **else**
15:             Upgrade Path.
16:         **end if**
17:     **end if**
18:     Neural identification
19:     Perform next control action $u_{(k)}$
20:     Get odometry.
21:     Get observations $z_{(k)}$.
22:     Perform prediction step of SLAM.
23:     Perform measurement update of SLAM.
24:     Add new features.
25: **end while**

The first lines in algorithm (2) are used to initialize robot pose x, covariance matrix of robot pose P, state map used in RL and observation z. After this, if there is an obstacle or a change in the previous information about the environment, the experience (policy) learned with RL is used to look for a new local path to try to overcome the obstacle. Navigation policy learned with RL is used to find new paths for most of the obstacles. This is one of the major advantages of our approach, it has the ability to re-plan the route

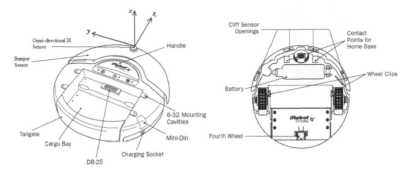

FIGURE 4.1: Vehicle specifications.

without having to recalculate or update travel cost of the sub-path that has already been explored before finding a change in the environment, this ability is acquired by the agent thanks to the RL algorithm that allows the agent to learn from experience, and this experience can be seen as a state-action memory. In the final lines of algorithm (2), the exploration of environment is continued and SLAM measures are taken in order to complete the map.

4.5 Experimental Results

The proposed schemes are carried out in a MATLAB and Simulink environment, applied on a differential robot model with sensors of vision and movement and sensor readings corrupted by noise, caused by wireless transmission. The anatomy and various components and body axes of the Quanser® are shown in Figure 4.1.

The Quanser Qbot[1] is an innovative autonomous ground robot system [104]. The vehicle comprises an iRobot Create® robotic platform, an array of optional infrared and sonar sensors, and a Logitech® Quickcam Pro 9000 USB camera. The diameter of the vehicle is 34 cm, and its height (without camera attachment) is 7 cm; It is driven by two differential drive wheels and comes with a bumper sensor and an omni-directional infrared receiver. The Quanser Controller Module (QCM) is an embedded system mounted on the vehicle, which uses the Gumstix computer to run QuaRC, Quanser's real-time control software [104] .

The Qbot[1] is accessible through three different block sets: the Roomba block set to drive the vehicle, the HIL block set to read from sensors and/or write to servo outputs, and finally the OpenCV block set to access the camera. The controllers are developed in Simulink® with QuaRC on the host computer thought wireless communication. This type of communication im-

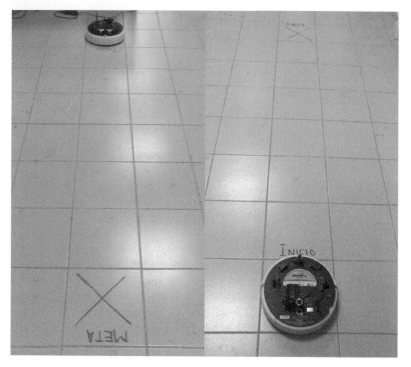

FIGURE 4.2: Starting position in the real environment.

plies noise, delays and uncertainties, which are absorbed by the neural network which is learning in real time the system behavior, and is capable of predicting the next state in order to avoid the lost/corrupted information.

It is important to mention that the communication with the robot is wireless, and this implies loss of information, noise, delays, uncertainties, attenuations, and fading, among other problems. Due to the nature of the neural network used, the algorithm is able to learn from the previous experience generating its own weight distributions on the links. This learning ability to organize the information causes the power to appropriately respond for data or situations to which it has not been previously exposed.

In this case a map of 33×19 pixels is used obtaining a total of 627 possible states and 5016 possible actions, each state represents $1/3$ m^2; however, the map can be of any desired size. Several experiments were conducted in order to show the system capability to complete its goal despite changing environments.

The experiments are performed with sampling time of 0.05 seconds, a range of 33 cm vision for SLAM, the ability to detect obstacles 33 cm away, a rate of 0.05 m/s and some positions blocked, then the tests performed were adapted to the workbench (Fig. 4.2), in which each state is represented by a block of 33×33 cm (0.1089 m^2).

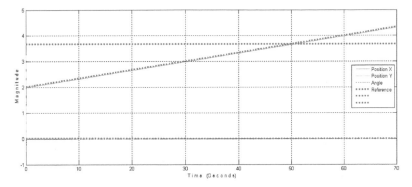

FIGURE 4.3: System and reference behavior, path without obstacles, positions (X, Y) meters, angle θ radians.

The tests were performed with a training map where obstacles were not present; the robot confronted an unknown environment in real time; the goal was to prove that the system is able to get sub-routes in real time to avoid the obstacles that were placed randomly.

Once the initial route has been obtained the navigation is started following the obtained path, but once an obstacle is encountered by the SLAM observations, it checks into the map state action and avoids the obstacle with decisions that come from the previous experience.

It is important to note that each optimal route is modified by the robot in order to avoid obstacles. There is a little displacement between the error in the pose estimation from SLAM and the noise that the control actions and data measurement have. The state-map is corrupted every time with noise in the route planning part; this noise is discarded and re-added every time the planning takes place. Thus, every time the system perceives a different map.

Figure 4.3 shows the development of the system and the reference in the first test, which was performed without obstructing the movement of the robot. As a result, the mobile agent is moved in a straight line without errors.

Note: The peaks on the graphics are caused by dynamic obstacles or unexpected changes in the environment. When this occurs, RL is used to correct the path with the learned optimal policy. Even though the changes are fast the RL and indirect control respond on real time to these changes as can be seen in the figures; also it represents when the robot turns.

For the other tests, the robot was blocked in different times and places; then the system was forced to generate alternative routes, and to be able to complete the goal, regardless of the obstacles placed (Fig. 4.4), lost of data, time delays and other problems already mentioned. The obtained map is depicted in Fig. 4.5.

The results obtained from the different tests between the model (reference)

FIGURE 4.4: Path with obstacles made in real time.

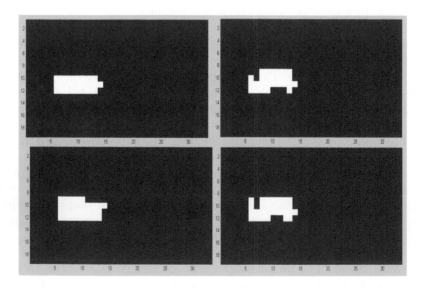

FIGURE 4.5: Maps.

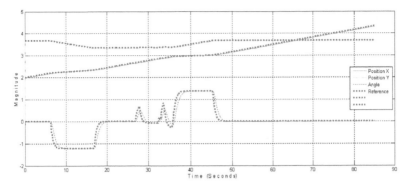

FIGURE 4.6: System and reference develop with some obstacles placed in real time.

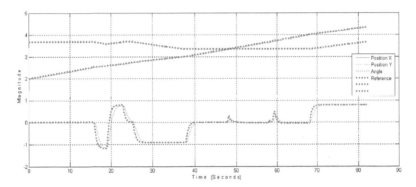

FIGURE 4.7: Reference system Behavior, with some obstacles placed in the path of the robot, Test 3.

and the system output for the X, Y position, and angle θ, are shown in Figures 4.6, 4.7 and 4.8, respectively.

Position errors in trajectory tracking can be seen in Figures 4.9, 4.10 and 4.11 for (X, Y, θ) respectively, and the generated map is shown in Figure 4.12.

Finally, the results of neuronal identification error performance for robot states (X, Y, θ) are shown in Figure 4.13, 4.14 and 4.15.

The following example was done with a complex map to see the robot's ability to complete the objective. The environment was modified to see the response and behavior of the robot to changes and obstacles in the environment.

Figure 4.16 shows the real environment for application, development with an initial position $(01, 02)$ and goal $(08, 13)$, (Y, X) respectively.

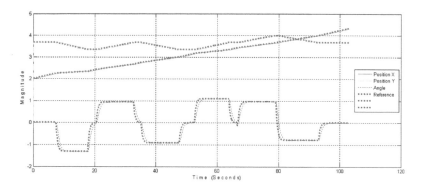

FIGURE 4.8: Reference system Behavior, with some obstacles placed in the path of the robot, Test 4.

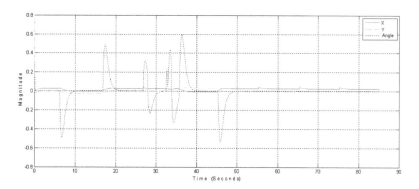

FIGURE 4.9: Error in the robot's position, reference-system, Test 2.

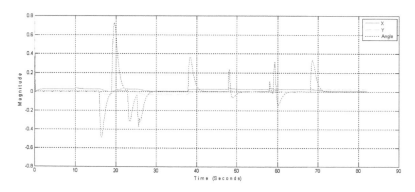

FIGURE 4.10: Position error, reference-system, Test 3.

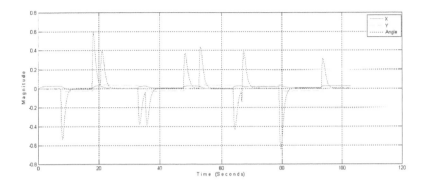

FIGURE 4.11: Position error, Test 4.

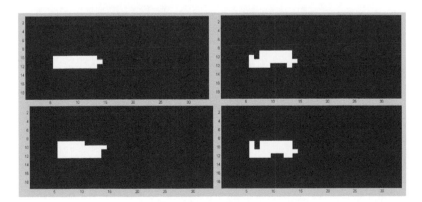

FIGURE 4.12: Maps created.

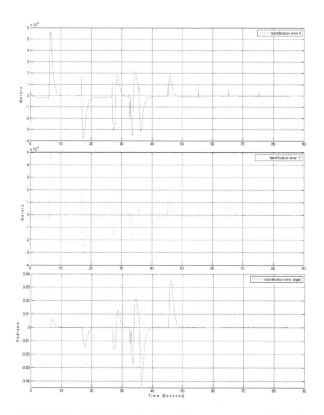

FIGURE 4.13: Neural identification error, Test 2.

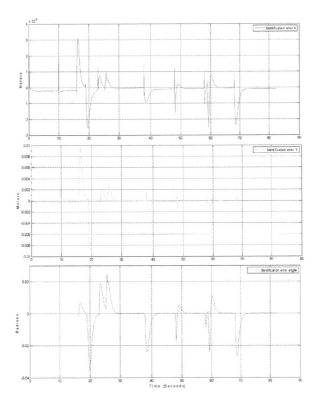

FIGURE 4.14: Neural identification error, Test 3.

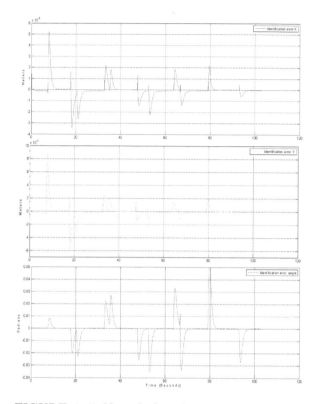

FIGURE 4.15: Neural identification error, Test 4.

FIGURE 4.16: Real environment.

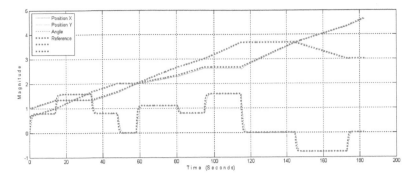

FIGURE 4.17: Development and the reference system in a more complicated map.

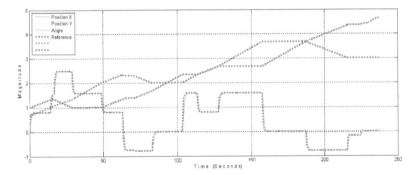

FIGURE 4.18: Development and the reference system, with obstacles placed.

Later Figures 4.17 and 4.18 show the results of each application in the environment with/without obstacles in .

And identification errors can be seen in Figures 4.19 and 4.20.

Figure 4.21 is the maps created by the application on the 2 different tests.

Besides all problems and disadvantages normally presented by wireless communication and noise sensors, the system is provided with drawbacks to see how the algorithm responds to these situations.

Figures 4.22 and 4.23 show how the system is able to continue without interrupting the execution with loss/corrupted information and recover in their positions (X, Y, θ) get back to the correct signal. Remember that the sampling time is 0.05 seconds so that the loss of information is large.

Figure 4.24 shows the rates of a test where the reference was cut by 3 seconds, and the system did not receive the signal reference on that time as can be seen in Figure 4.25. In response, the system decreased the speed waiting for the reference to be restored; at the time the reference is recovered

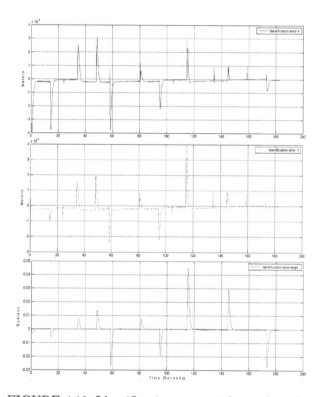

FIGURE 4.19: Identification error without obstacles.

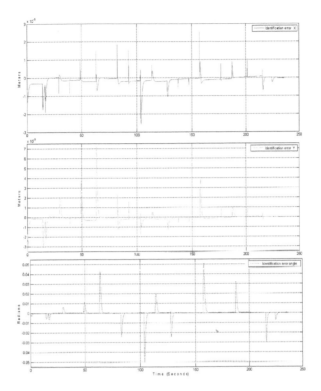

FIGURE 4.20: Identification error with obstacles.

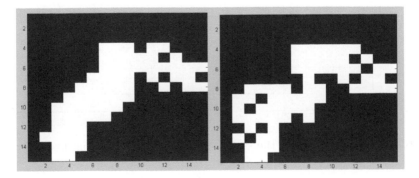

FIGURE 4.21: Maps created.

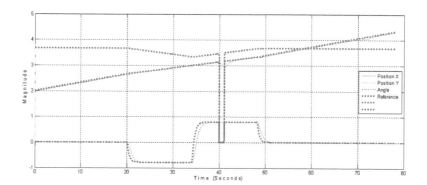

FIGURE 4.22: System response to loss/incorrect information at second 40.

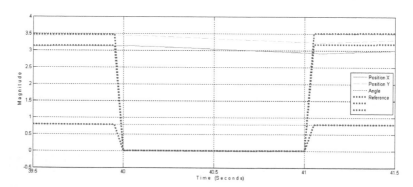

FIGURE 4.23: Details of the system's response to erroneous/lost information

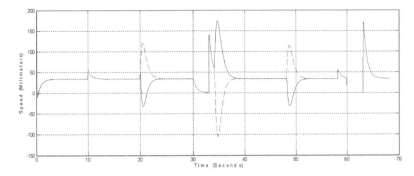

FIGURE 4.24: Speeds applied to the robot on a test where the reference was cut.

the speed of the robot is restored to bring the robot to the given reference (Figure 4.26).

Following at second 60, the same test was performed but in this case the driver was interrupted (Figure 4.27), as the driver was not working the law control high (Figure 4.28) in order to compensate for the error position and tried to take the robot to the reference, but when the control was applied again the controller returned to the normal speed.

Figures 4.29 and 4.30 show the case of delays, where the reference is delayed 1 second on time 40. In response, the algorithm again slows the robot speed until receiving signal again; however, in this case it does not make a jump on speed since there was no loss of information, just outdated information.

Figure 4.31 displays how the error decreases while the reference is delayed, this happens because there is an induced time-delay to the controller in order to reach the final reference, but when the signal returns, the error increases, because the reference is a step forward and there will be an error resulting in a time minimum k,

As can be seen in the different cases, the robot is able to reach the goal regardless of the obstacles, noise, delay or uncertainty. There are some errors in position, but the results are still the expected.

4.6 Conclusions

In this chapter, an artificial intelligence-based system for the navigation and exploration task of a mobile robot, is presented and tested. A robot system that is capable of navigating unknown environments even with uncertainties in the model and robot or in the environment has been developed. This can

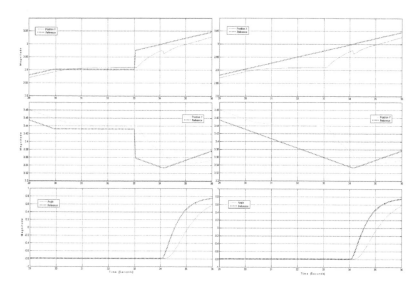

FIGURE 4.25: Detail of the System-Reference on time 30-33. The first column shows the reference (X, Y, θ) with the interruption, and the second column the real signal to follow, which was interrupted.

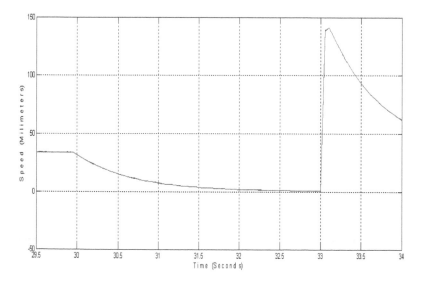

FIGURE 4.26: Detail of the velocity: from second 30 to 33 no reference for (X, Y, θ) was sent, and in response the system decreased the speed until the reference continued.

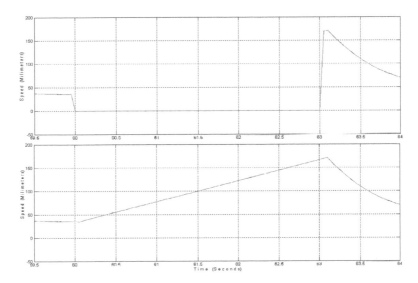

FIGURE 4.27: Detail of the velocity: from time 60 to 63 the control was stopped, and the first figure shows the system behavior when the driver stopped work, and the second shows what the system tried to do to respond at the problem.

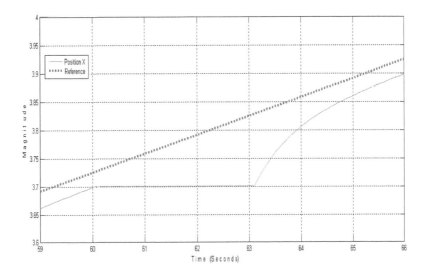

FIGURE 4.28: Detail of the System-Reference. X position on time 60-63 when the control stopped.

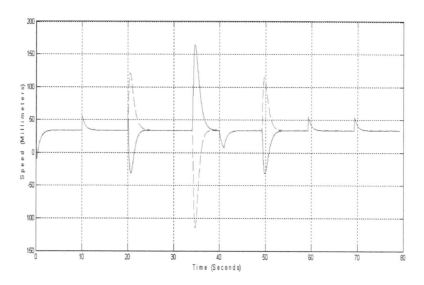

FIGURE 4.29: Wheel speeds with delay on the reference at time 40.

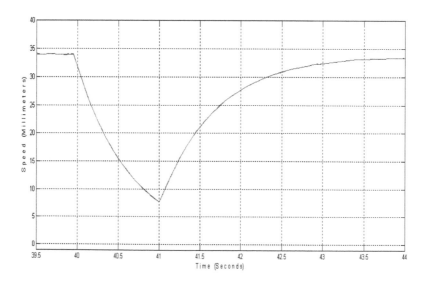

FIGURE 4.30: Detail on speed reduction.

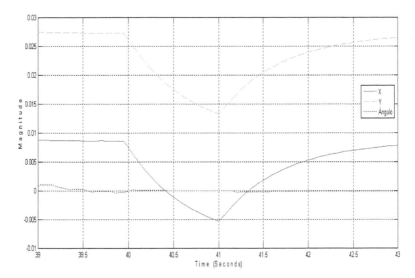

FIGURE 4.31: Error signal delay.

be attained because a RHONN structure is used to design a neural identifier which is flexible and robust to noise. In spite of wireless communication which implies loss of information, the results show the effectiveness of the proposed schemes. In addition, the qualities of RL are added to the algorithm to obtain a robust system, capable of handling unknown and long state dynamic noisy environments.

5

Holonomic Robot Control Using Neural Networks

CONTENTS

5.1 Introduction

The greatest success in robot automation has been achieved in the manufacturing industry. The most popular robot in such industry is the manipulator robot. These robots have to be installed at some point during the production line. They can perform repetitive tasks with great speed and accuracy, Fig. 5.1.

Although manipulator robots have great success in the manufacturing industry, these robots suffer from an important disadvantage, that is the lack of mobility. Manipulator robots can solve many tasks, however they can not move in their environment. Thus, its range of motion is limited.

In contrast, mobile robots do not suffer from a limited range of motion. These robots can move throughout the manufacturing factory, and are able to solve problems where they are needed, Fig. 5.2. Mobile robots have a lot of advantages; however, there are some important tasks that must be solved like robot localization, motion planning, visual servoing, obstacle avoidance, etc.

There are many types of mobile robots, which basically differ in the type of locomotion they have. Mobile robots can be divided into holonomic and non-

FIGURE 5.1: Robot manipulator.

holonomic robots. A holonomic robot has the same controllable DOF (degrees of freedom) than the total DOF of the robot, and in contrast a nonholonomic robot has less controllable DOF than the total DOF of the robot. In this case we focus in holonomic robots, which have the ability to move sideways.

A holonomic or omnidirectional robot is able to travel in every direction under any orientation. The term omnidirectional robot is used to describe the ability of a system to move instantaneously in any direction from any configuration [34].

In general, there are two types of omnidirectional wheeled platforms. The first one uses special wheels, and the other one includes conventional wheels. Special wheels are called Mecanum or Swedish wheels, which have been mostly studied for the omnidirectional mobile platforms (see Fig. 5.3). Conventional wheels can be divided into two types, caster wheels and steering wheels. Four Swedish wheels provides omnidirectional movement for a vehicle without needing a conventional steering system [34] [90]. In our case we use a Youbot robot, which has four Swedish wheels, and therefore the robot is able to move sideways.

In real situations the control based on the plant model could not perform as desired. This is due to external and internal disturbances, unmodelled dynamics, or uncertain parameters [46]. Given these problems, a more robust approach is required. In this case we use a recurrent high order neural network (RHONN) in order to identify the dynamics of the plant to be controlled. The advantages of the RHONN are its robustness, and its capacity to adjust its parameters in real time [114], [117], in addition to the incorporation of a priori system structure information [99]. In this case, the RHONN is used to identify each subsystem, under the assumption that the full state is available

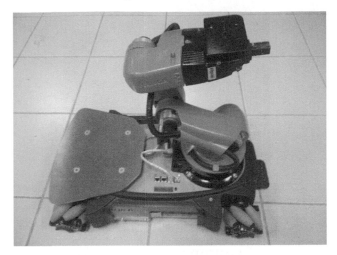

FIGURE 5.2: Mobile robot with manipulator.

for measurement. The learning algorithm for the local RHONN is implemented using an Extended Kalman Filter (EKF).

Nowadays, most of the technological problems involve complex large scale systems. To solve this type of problem it is recommended to avoid a centralized controller, the main reasons being due to the complexity of on-line computations, the reliability of the controller, the complexity of the controller design, and the implementation cost. In such cases a decentralized approach appears as an interesting option to take into account, [58].

In recent years, the decentralized stabilization of large-scale systems has received considerable attention [47, 60, 63, 69, 72, 74, 80]. The decentralized control can be defined as a system composed of several subsystems or agents, each one with a local control. In this approach a local control station directly considers the state of its subsystem, without considering the state of the other subsystems. The advantage of this approach, is that the subsystem controllers are simple, even though that the whole system is a complex large scale system [62],[85].

To determine the control signals that force a process to satisfy physical constraints and at the same time to minimize a performance criterion the optimal control theory is used [65]. Unfortunately, for nonlinear systems, it is required to solve the associated Hamilton Jacobi Bellman (HJB) equation, which is not an easy task.

To avoid the solution of the HJB equation an approach called inverse optimal control can be used [68]. For the inverse approach, a stabilizing feedback control law is developed and then it is established that this control law optimizes a cost functional. The main characteristic of the inverse approach is

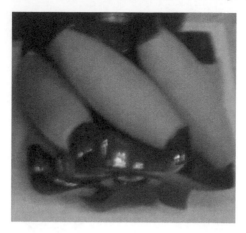

FIGURE 5.3: Swedish wheel.

that the cost functional is a posterior determined for the stabilizing feedback control law [32, 38, 100].

This chapter presents a discrete time neural control of an omnidirectional mobile robot with visual feedback. The approach consists of synthesizing a suitable controller for each subsystem. First, the dynamics of each subsystem is approximated by an identifier using a discrete-time recurrent high order neural network, trained with an extended Kalman filter algorithm. Then based on this neural model an inverse optimal controller is synthesized to avoid solving the Hamilton Jacobi Bellman (HJB) equation. The desired trajectory of the robot is computed during navigation using a camera sensor. Simulation results are presented to illustrate the effectiveness of the proposed control scheme.

The remainder of this paper is organized as follows: Section 5.2 presents the general inverse optimal control. Section 5.3 establishes the decentralized inverse optimal control. Section 5.4 establishes the discrete-time neural control. Section 5.5 describes the use of visual feedback to design a desired trajectory reference for the controller. The visual data are acquired from a camera mounted on the robot. Using visual data the controller drives the omnidirectional robot from its current pose toward a desired pose. Section 5.6 illustrates the simulation results of our proposed scheme. Finally, in Section 5.7 the conclusions are given.

5.2 Optimal Control

This section briefly gives details about optimal control methodology and its limitations. Let consider the discrete-time affine-in-the-input nonlinear system:

$$\chi_{k+1} = f(\chi_k) + g(\chi_k)u_k, \quad \chi(0) = \chi_0 \tag{5.1}$$

where $\chi_k \in \mathbb{R}^n$ is the state of the system, $u_k \in \mathbb{R}^m$ is the control input, $f(\chi_k) : \mathbb{R}^n \to \mathbb{R}^n$ and $g(\chi_k) : \mathbb{R}^n \to \mathbb{R}^{n \times m}$ are smooth maps, the subscript $k \in \mathbb{Z}^+ \cup 0 = \{0, 1, 2, \ldots\}$ stands for the value of the functions and/or variables at time k, $f(0) = 0$ and $rank\{g(\chi_k)\} = m \ \forall \chi_k \neq 0$. For system 5.1, it is desired to determine a control law u_k which minimizes a cost functional. The following cost function is associated with a trajectory tracking for the system 5.1

$$\mathcal{J}(z_k) = \sum_{n=k}^{\infty} (l(z_n) + u_n^T R u_n) \tag{5.2}$$

where $z_k = \chi_k - \chi_{\delta,k}$ with $\chi_{\delta,k}$ as the desired trajectory for χ_k; $z_k \in \mathbb{R}^n$; $\mathcal{J}(z_k) : \mathbb{R}^n \to \mathbb{R}^+$; $l(z_k) : \mathbb{R}^n \to \mathbb{R}^+$ is a positive semidefinite function and $R : \mathbb{R}^n \to \mathbb{R}^{m \times m}$ is a real symmetric positive definite weighting matrix. The meaningful cost functional (5.2) is a performance measure [65]. The entries of R may be functions of the system state in order to vary the weighting on control efforts according to the state value [65]. Considering the state feedback control approach, we assume that the full state χ_k is available.

From Bellman's optimality principle [12, 71], it is known that, for the infinite horizon optimization case, the value function $V(z_k)$ becomes time invariant and satisfies the discrete-time (DT) Bellman equation [2, 12, 97]

$$V(z_k) = \min_{u_k} \{l(z_k) + u_k^T R u_k + V(z_{k+1})\} \tag{5.3}$$

where $V(z_{k+1})$ depends on both z_k and u_k by means of z_{k+1} in (5.1). Note that the DT Bellman equation is solved backward in time [2]. In order to establish the conditions that the optimal control law must satisfy, we define the discrete-time Hamiltonian $\mathcal{H}(z_k, u_k)$ [49] as

$$\mathcal{H}(z_k, u_k) = l(z_k) + u_k^T R u_k + V(z_{k+1}) - V(z_k). \tag{5.4}$$

The Hamiltonian is a method of adjoining constraint (5.1) to the performance index (5.2), and then, solving the optimal control problem by minimizing the Hamiltonian without constraints [71].

The optimal control law is formulated as

$$u_k^* = -\frac{1}{2} R^{-1} g^T(\chi_k) \frac{\partial V(z_{k+1})}{\partial z_{k+1}} \tag{5.5}$$

with the boundary condition $V(0) = 0$; u_k^* is used when we want to emphasize that u_k is optimal. The discrete-time Hamilton-Jacobi-Bellman (HJB) equation is described by

$$
\begin{aligned}
0 \;=\;& l(z_k) + V(z_{k+1}) - V(z_k) \\
+\;& \frac{1}{4}\frac{\partial V^T(z_{k+1})}{\partial z_{k+1}} g(\chi_k) R^{-1} g^T(\chi_k)\frac{\partial V(z_{k+1})}{\partial z_{k+1}}
\end{aligned}
\tag{5.6}
$$

The solution of the HJB partial-differential Eq. (5.6) is not straightforward; this is one of the main disadvantages of the discrete-time optimal control for nonlinear systems. To overcome this problem, we propose the use of inverse optimal control.

5.3　Inverse Optimal Control

Given a class of disturbed discrete-time nonlinear and interconnected systems

$$
\chi_{i,k+1}^{j} = f_i^{j}(\chi_{i,k}^{j}) + g_i^{j}(\chi_{i,k}^{j})u_{i,k} + \Gamma_{i\ell,k}^{j}(\chi_\ell)
\tag{5.7}
$$

where $i = 1,\dots,\gamma$, $j = 1,\dots,n_i$, $\chi_i \in \mathbb{R}^{n_i}$, $\chi_i = [\chi_i^{1T}\chi_i^{2T}\dots\chi_i^{rT}]^T$, $\chi_i^{j} \in \mathbb{R}^{n_{ij}\times 1}$ $u_i \in \mathbb{R}^{m_i}$; γ is the number of subsystems, χ_ℓ reflects the interaction between the i-th and the ℓ-th subsystem (agent) with $1 \le \ell \le \gamma$. We assume that f_i, B_i and Γ_i are smooth and bounded functions, $f_i^{j}(0) = 0$ and $B_i^{j}(0) = 0$. Without loss of generality, $\chi = 0$ is an equilibrium point of (5.7), which is to be used later.

For the inverse optimal control approach, let us consider the discrete-time affine-in-the-input nonlinear system:

$$
\chi_{i,k+1} = f_i(\chi_{i,k}) + g_i(\chi_{i,k})u_{i,k}, \quad \chi_{i,0} = \chi_i(0)
\tag{5.8}
$$

with $i = 1,\dots,\gamma$; γ is the number of subsystems (agents). Where $\chi_{i,k} \in \mathbb{R}^{n_i}$ are the states of the systems, $u_{i,k} \in \mathbb{R}^{m_i}$ are the control inputs, $f_i(\chi_{i,k})$: $\mathbb{R}^{n_i} \to \mathbb{R}^{n_i}$ and $g(\chi_k) : \mathbb{R}^{n_i} \to \mathbb{R}^{n_i\times m_i}$ are smooth maps, the subscript $k \in \mathbb{Z}^+\cup 0 = \{0,1,2,\dots\}$ will stand for the value of the functions and/or variables at the time k. We establish the following assumptions and definitions which allow the inverse optimal control solution via the CLF approach.

Assumption 1 *The full state of system (5.8) is measurable.*

Definition 1 ([118]) *Consider the tracking error $z_{i,k} = \chi_{i,k} - \chi_{i\delta,k}$ with $\chi_{i\delta,k}$ as the desired trajectory for $\chi_{i,k}$. Let define the control law*

$$
u_{i,k}^{*} = -\frac{1}{2}R_i^{-1}g_i^T(\chi_{i,k})\frac{\partial V_i(z_{i,k+1})}{\partial z_{i,k+1}}
\tag{5.9}
$$

which will be inverse optimal stabilizing along the desired trajectory $\chi_{i\delta,k}$ if:

- *(i) it achieves (global) asymptotic stability for system (5.8) along reference $\chi_{i\delta,k}$;*

- *(ii) $V_i(z_{i,k})$ is (radially unbounded) positive definite function such that inequality*

$$\overline{V}_i := V_i(z_{i,k+1}) - V_i(z_{i,k}) + u_{i,k}^{*T} R_i u_{i,k}^* \leq 0 \tag{5.10}$$

is satisfied.

When $l_i(z_{i,k}) := -\overline{V}_i \leq 0$ is selected, then $V_i(z_{i,k})$ is a solution for the HJB equation

$$\begin{aligned} 0 &= l_i(z_{i,k}) + V_i(z_{i,k+1}) - V_i(z_{i,k}) \\ &+ \frac{1}{4} \frac{\partial V_i^T(z_{i,k+1})}{\partial z_{i,k+1}} g_i(\chi_{i,k}) R^{-1} g_i^T(\chi_{i,k}) \frac{\partial V_i(z_{i,k+1})}{\partial z_{i,k+1}} \end{aligned} \tag{5.11}$$

and the cost functional (5.2) is minimized. It is possible to establish the main conceptual differences between optimal control and inverse optimal control as follows:

- For optimal control, the meaningful cost indexes $l_i(z_{i,k}) \leq 0$ and $R_i > 0$ are given a priori; then, they are used to calculate $u_i(z_{i,k})$ and $V_i(z_{i,k})$ by means of the HJB equation solution.

- For inverse optimal control, a candidate CLF ($V_i(z_{i,k})$) and the meaningful cost index R_i are given a priori, and then these functions are used to calculate the inverse control law $u_i^*(z_{i,k})$ and the meaningful cost index $l_i(z_{i,k})$, defined as $l_i(z_i) := -\overline{V}_i$.

As established in Definition 1, the inverse optimal control law for trajectory tracking is based on the knowledge of $V_i(z_{i,k})$. Thus, we propose a CLF $V_i(z_{i,k})$, such that (i) and (ii) are guaranteed. That is, instead of solving (5.6) for $V_i(z_{i,k})$, a quadratic CLF candidate $V_i(z_{i,k})$ is proposed with the form:

$$V_i(z_{i,k}) = \frac{1}{2} z_{i,k}^T P_i z_{i,k}, \quad P_i = P_i^T > 0 \tag{5.12}$$

for control law (5.9) in order to ensure stability of the tracking error $z_{i,k}$, where

$$\begin{aligned} z_{i,k} &= \chi_{i,k} - \chi_{i\delta,k} \tag{5.13} \\ &= \begin{bmatrix} (\chi_{i,k}^1 - \chi_{i\delta,k}^1) \\ (\chi_{i,k}^2 - \chi_{i\delta,k}^2) \\ \vdots \\ (\chi_{i,k}^n - \chi_{i\delta,k}^n) \end{bmatrix} \tag{5.14} \end{aligned}$$

Moreover, it will be established that control law (5.9) with (5.12), which is referred to as the inverse optimal control law, optimizes a cost functional of the form (5.2).

Consequently, by considering $V_i(\chi_{i,k})$ as in (5.12), control law (5.9) takes the following form:

$$
\begin{aligned}
\alpha_i(\chi_{i,k}) := u_{i,k}^* &= -\frac{1}{4} R_i g_i^T(\chi_{i,k}) \frac{\partial z_{i,k+1}^T P_i z_{i,k+1}}{\partial z_{i,k+1}} \\
&= -\frac{1}{2} R_i g_i^T(\chi_{i,k}) P_i z_{i,k+1} \\
&= -\frac{1}{2} (R_i(\chi_{i,k}) + \frac{1}{2} g_i^T(\chi_{i,k}) P_i g_i(\chi_{i,k}))^{-1} \\
&\quad \times g_i^T(\chi_{i,k}) P_i(f_i(\chi_{i,k}) - \chi_{i\delta,k+1})
\end{aligned}
\tag{5.15}
$$

It is worth pointing out that P_i and R_i are positive definite and symmetric matrices; thus, the existence of the inverse in (5.15) is ensured.

Once we have proposed a CLF for solving the inverse optimal control in accordance with Definition 1, the respective solution is presented, for which P_i is considered a fixed matrix.

Lemma 2 *Consider the affine discrete-time nonlinear system (5.8) with $i = 1$. If there exists a matrix $P_i = P_i^T > 0$ such that the following inequality holds:*

$$
\begin{aligned}
\frac{1}{2} f_i^T(\chi_{i,k}) P_i f_i(\chi_{i,k}) &+ \frac{1}{2} \chi_{i\delta,k+1}^T P_i \chi_{i\delta,k+1} - \chi_{i,k}^T P_i \chi_{i\delta,k}^T \\
- \frac{1}{2} \chi_{i\delta,k}^T P_i \chi_{i\delta,k} &- \frac{1}{4} P_{i1}^T(\chi_{i,k}, \chi_{i\delta,k})(R_i + P_{i2}(\chi_{i,k}))^{-1} \\
\times P_{i1}(\chi_{i,k}, \chi_{i\delta,k}) & \\
&\leq -\frac{1}{2} \|P_i\| \|f_i(\chi_{i,k})\|^2 - \frac{1}{2} \|P_i\| \|\chi_{i\delta,k+1}\|^2 \\
&\quad - \frac{1}{2} \|P_i\| \|\chi_{i,k}\|^2 - \frac{1}{2} \|P_i\| \|\chi_{i\delta,k}\|^2
\end{aligned}
\tag{5.16}
$$

where $P_{i,1}(\chi_{i,k}, \chi_{i\delta,k})$ and $P_{i,2}(\chi_{i,k})$ are defined as

$$
P_{i,1}(\chi_{i,k}, \chi_{i\delta,k}) = g_i^T(\chi_{i,k}) P_i(f_i(\chi_{i,k}) - \chi_{i\delta,k+1})
\tag{5.17}
$$

and

$$
P_{i,2}(\chi_{i,k}) = \frac{1}{2} g_i^T(\chi_{i,k}) P_i g_i(\chi_{i,k})
\tag{5.18}
$$

respectively, then system (5.8) with control law (5.15) guarantees asymptotic trajectory tracking along the desired trajectory $\chi_{i\delta,k}$, where $z_{i,k+1} = \chi_{i,k+1} - \chi_{i\delta,k+1}$.

Moreover, with (5.12) as a CLF, this control law is inverse optimal in the sense that it minimizes the meaningful functional given by

$$\mathcal{J}_i(z_{i,k}) = \sum_{k=0}^{\infty} (l_i(z_{i,k}) + u_{i,k}^T R_i(z_{i,k}) u_{i,k}) \tag{5.19}$$

with

$$l_i(z_{i,k}) = -\overline{V}_i|_{u_{i,k}^* = \alpha_i(z_{i,k})} \tag{5.20}$$

and optimal value function $\mathcal{J}_i(z_{i,k}) = V_i(z_0)$.

This lemma is adapted from [118] for each isolated subsystem, which allows establishing the following theorem.

5.4 Holonomic Robot

In this work we consider a robot with four Swedish wheels with rollers at $45°$, each driven by a separate motor as shown in Fig. 5.4

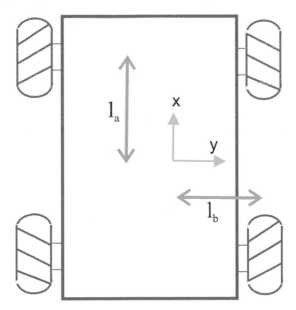

FIGURE 5.4: Omnidirectional mobile robot.

5.4.1 Motor dynamics

The dynamics of the DC motors of the robot can be expressed in the following state-space model [23]:

TABLE 5.1: State-space model for the dynamicos of the DC motors.

$R_a = 0.6\Omega$	$K_b = 0.8\frac{Vs}{rad}$
$L_a = 0.012H$	$J = 0.0167\frac{kgm^2}{s^2}$
$K_t = 0.8\frac{Nm}{A}$	$b = 0.0167Nms$

$$
\begin{aligned}
\chi^1_{i,k+1} &= \chi^1_{i,k} + T(-\frac{b}{J}\chi^1_{i,k} + \frac{K_t}{J}\chi^2_{i,k}) \\
&\quad + \sum_{\ell=1,\ell\neq i}^{\gamma} \Gamma_{1\ell,k}(\chi_\ell) \\
\chi^2_{i,k+1} &= \chi^2_{i,k} + T(-\frac{K_b}{L_a}\chi^1_{i,k} - \frac{R_a}{L_a}\chi^2_{i,k} + \frac{1}{L_a}u_{i,k}) \\
&\quad + \sum_{\ell=1,\ell\neq i}^{\gamma} \Gamma_{2\ell,k}(\chi_\ell)
\end{aligned}
\tag{5.21}
$$

where χ^1_i represent the angular velocity in $\frac{rad}{s}$ for each motor respectively with $i = 1\ldots 4$. Accordingly, each actuated wheel is considered as a subsystem. χ^2_i is the armature current in Amp. $\Gamma_{i\ell}(\chi_\ell)$ reflect the interaction between the i-th and the ℓ-th subsystem with $1 \leq \ell \leq \gamma$. The input terminal voltage u_i is taken to be the controlling variable. R_a and L_a are the armature inductance in H and resistance in ohm respectively. K_t is the torque factor constant in $\frac{Nm}{Amp}$. K_b is the back emf constant in $\frac{Vs}{rad}$. J represents the moment of inertia in $\frac{kgm^2}{s^2}$ and b is the coefficient of viscous friction which opposes the direction of motion in Nms. The sampling step is $T = 0.01$.

5.4.2 Neural identification design

We apply the neural identifier, developed in [80, 117], to obtain a discrete-time neural model for the electrically driven Youbot robot (5.22) which is trained with the EKF respectively, as follows:

$$
\begin{aligned}
x^1_{i,k+1} &= w_{11i,k}S(\chi^1_{i,k}) + w_{12i,k}S(\chi^1_{i,k}) \\
&\quad + w'_{1i}\chi_{2i,k} \\
x^2_{i,k+1} &= w_{21i,k}S(\chi^2_{i,k}) + w_{22i,k}S(\chi^1_{i,k}) + w_{23i,k}S(\chi^2_{i,k}) \\
&\quad + w'_{2i}u_{i,k}
\end{aligned}
\tag{5.22}
$$

where x^1_i and x^2_i identify the angular velocities χ^1_i and the motor currents χ^2_i, respectively. The NN training is performed on-line, and all of its states are initialized, randomly. The RHONN parameters are heuristically selected as:

$$P_{iq}^1(0) = 1 \times 10^{10} I \quad R_{iq}^1(0) = 1 \times 10^8 \quad w'_{1i} = 1$$
$$P_{iq}^2(0) = 1 \times 10^{10} I \quad R_{iq}^2(0) = 5 \times 10^3 \quad w'_{2i} = 1$$
$$Q_{iq}^1(0) = 1 \times 10^7 I \quad Q_{iq}^2(0) = 5 \times 10^3 I$$

where I is the identity matrix. It is important to consider that for the EKF-learning algorithm the covariances are used as design parameters [38], [52]. The neural network structure (5.22) is determined heuristically in order to minimize the state estimation error. It is worth noting that (5.22) constitutes a series-parallel identifier [39, 117] and [59] does not consider explicitly the interconnection terms, whose effects are compensated by the neural network weights update.

5.4.3   Control design

The goal is to force the state $x_{i,k}^1$ to track a desired reference signal $\chi_{i\delta,k}^1$, which is achieved by a control law as described in Section 5.3. First the tracking error is defined as

$$z_{i,k}^1 = x_{i,k}^1 - \chi_{i\delta,k}^1$$

Then using (5.22) and introducing the desired dynamics for $z_{i,k}^1$ results in

$$
\begin{aligned}
z_{i,k+1}^1 &= w_{1i,k}\varphi_1(\chi_{i,k}^1) + w'_{1i,k}\chi_{i,k}^2 - \chi_{i\delta,k+1}^1 \\
&= K_i^1 z_{1i,k}
\end{aligned}
\tag{5.23}
$$

where $|K_i^1| < 1$. The desired value $\chi_{i\delta,k}^2$ for the pseudo-control input $\chi_{i,k}^2$ is calculated from (5.23) as

$$
\begin{aligned}
\chi_{i\delta,k}^2 = (w'_{1i,k})^{-1}(&-w_{1i,k}\varphi_{1i}(\chi_{i,k}^1) \\
&+\chi_{i\delta,k+1}^1 + K_i^1 z_{i,k}^1)
\end{aligned}
\tag{5.24}
$$

At the second step, we introduce a new variable as

$$z_{i_k}^2 = x_{i,k}^2 - \chi_{i\delta,k}^2$$

Taking one step ahead, we have

$$
\begin{aligned}
z_{i,k+1}^2 = w_{2i,k}\varphi_{2i}(\chi_{i,k}^1, \chi_{i,k}^2) &+ w'_{2i,k}u_{i,k} \\
&-\chi_{i\delta,k+1}^2
\end{aligned}
\tag{5.25}
$$

where $u_{i,k}$ is defined as

$$
\begin{aligned}
u_{i,k} = -\frac{1}{2}\left(R_i(z_k) + g_i^T(x_{i,k})P_i g_i(z_k)\right)^{-1} \\
\times g_i^T(x_k)P_i(f_i(x_{i,k}) - x_{i\delta,k+1})
\end{aligned}
\tag{5.26}
$$

and the controllers parameters are shown below:

$$P_i = \begin{bmatrix} 1.6577 & 0.6299 \\ 0.6299 & 2.8701 \end{bmatrix}$$

Remark 3 *By means of stability analysis of the proposed scheme's different components, it is possible to establish tracking of a desired trajectory x_{id}^j defined in terms of the plant state χ_i^j, due to the separation principle for discrete-time nonlinear systems [73].*

5.4.4   Omnidirectional mobile robot kinematics

The omnidirectional mobile robot can spin around its vertical axis and move in any trajectory in the plane by varying the relative speed and the direction of rotation of the four wheels; see Fig. 5.4.

The kinematic model of the mobile robot is

$$\begin{bmatrix} x_{1\delta,k+1}^1 \\ x_{2\delta,k+1}^1 \\ x_{3\delta,k+1}^1 \\ x_{4\delta,k+1}^1 \end{bmatrix} = \frac{1}{R} \begin{bmatrix} -1 & 1 & (l_a + l_b) \\ 1 & 1 & -(l_a + l_b) \\ -1 & 1 & -(l_a + l_b) \\ 1 & 1 & (l_a + l_b) \end{bmatrix} \begin{bmatrix} v_{x,k} \\ v_{y,k} \\ \omega_k \end{bmatrix} \tag{5.27}$$

where the inputs $v_{x,k}$, $v_{y,k}$ and ω_k represent the driving velocity and the steering velocity respectively [90]. The value R denotes the radius of the wheel, the values l_a and l_b are defined as in Fig. 5.4.

5.5   Visual Feedback

The use of visual feedback to control a robot is commonly termed *visual servoing* or *visual control* [24, 25, 37, 57]. In this work we assume that a monocular camera is mounted directly on the mobile robot, in which case motion of the robot induces camera motion.

The objective of the visual feedback is the minimization of

$$e(t) = s(t) - s^* \tag{5.28}$$

where $s(t)$ denote the features extracted from the current pose, and s^* denote the features extracted from the desired pose.

During the motion of the robot the camera suffers a motion $\mathbf{v}_c = (v_c, \omega_c)$, where v_c is the linear velocity and ω is the angular velocity. The relationship between this velocity and $\dot{s}$ is

$$\dot{s} = \mathbf{L}\mathbf{v}_c \tag{5.29}$$

The relationship between the camera velocity and the time variation error can be determined with (6.21) and (5.29), that is

$$\dot{e} = \mathbf{L}\mathbf{v}_c \tag{5.30}$$

Where the interaction matrix $\mathbf{L}$ can be defined as [24]

$$\mathbf{L}_x = \begin{bmatrix} -\frac{1}{Z} & 0 & \frac{x}{Z} & xy & -(1+x^2) & y \\ 0 & -\frac{1}{Z} & \frac{y}{Z} & 1+y^2 & -xy & -x \end{bmatrix}$$

The Z value represents the depth of the feature relative to the camera frame.

The mobile robot has only 3 DOF that is v_x, v_y, ω_z, forward velocity, lateral velocity and angular velocity respectively. Since the robot has only 3 DOF we can rewrite as

$$\mathbf{L}_x = \begin{bmatrix} \frac{-1}{Z} & 0 & y \\ 0 & \frac{-1}{Z} & -x \end{bmatrix} \tag{5.31}$$

We can note from (5.5) that each feature point provides two equations, therefore, for a 3 DOF problem we need at least two features.

In the mobile robot case, the matrix L can be defined as

$$\mathbf{L} = \mathbf{L}_x \, {}^c\mathbf{V}_r \, {}^r\mathbf{J}_r \tag{5.32}$$

The matrix ${}^c\mathbf{V}_r$ is a motion transformation matrix defined as [101]

$$ {}^c\mathbf{V}_r = \begin{bmatrix} \mathbf{R} & [t]_\times \mathbf{R} \\ \mathbf{0}_{3\times3} & \mathbf{R} \end{bmatrix} \tag{5.33}$$

where $\mathbf{R}$ is the rotation matrix that relates the camera and robot frameworks, and $[t]_\times$ represents the skew symmetric matrix associated with the vector t. The matrix ${}^r\mathbf{J}_r$ is defined as

$$ {}^r\mathbf{J}_r = \begin{bmatrix} 1 & 0 & 0 \\ 0 & 1 & 0 \\ 0 & 0 & 0 \\ 0 & 0 & 0 \\ 0 & 0 & 0 \\ 0 & 0 & 1 \end{bmatrix} \tag{5.34}$$

From (5.30) and (5.32) we can define the velocity input to the robot as [24]

$$\mathbf{v}_r = -\lambda \mathbf{L}^+ \mathbf{e} \tag{5.35}$$

where $\mathbf{L}^+$ is the pseudo-inverse of the matrix $\mathbf{L}$. The features of the error (6.21) are defined as $s = (x \ \log(Z))^\top$ and $s^* = (x^* \ \log^*(Z))^\top$, and note that $\log(Z)$ and $\log^*(Z)$ is a supplementary normalized coordinate [86].

Equation 5.35 is used to estimate the desired velocity of the robot. Then, the desired velocity of each motor is computed with 5.27. Finally, the velocity of each motor is used for the reference for the neural network controller.

5.6   Simulation

In this section we present the simulations results. The initial pose of the robot was $x = -1, y = 0.2, \theta = 5°$, and the desired pose was $x = 0, y = 0, \theta = 0$. The velocities of the robot are computed using (5.35). The motor speeds are computed using (5.27), which is the input of the neural network.

The linear velocity is shown in Fig. 5.5, and the angular velocity is shown in Fig. 5.6. The neural network will control the motors of the omnidirectional mobile robot in order to achieve the desired velocities. It is important to note that the neural network is able to compensate for the inaccuracies in the model of the robot.

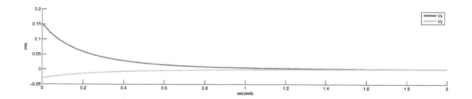

FIGURE 5.5: Linear velocities of the omnidirectional robot.

The results are shown as follows: Fig. 5.7 to Fig. 5.10 present angular velocity trajectory tracking for each i-th motor on the top, and the current trajectory tracking error for each i-th motor at the bottom.

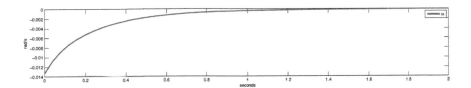

FIGURE 5.6: Angular velocity of the omnidirectional robot.

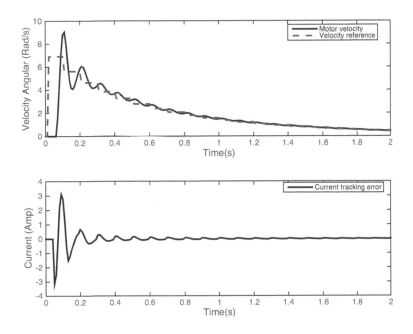

FIGURE 5.7: Trajectory tracking of the front right motor.

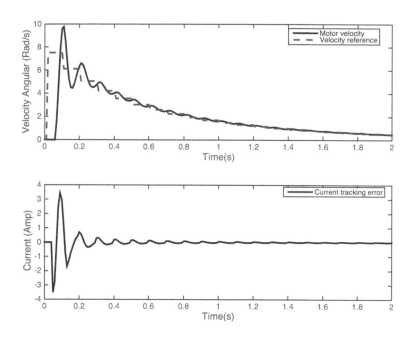

FIGURE 5.8: Trajectory tracking of the front left motor.

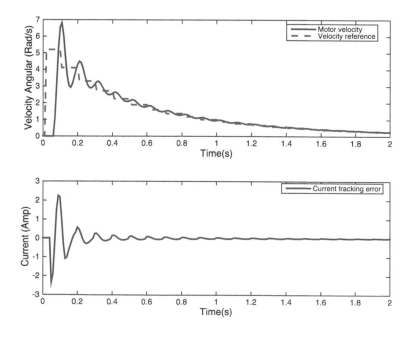

FIGURE 5.9: Trajectory tracking of the back right motor.

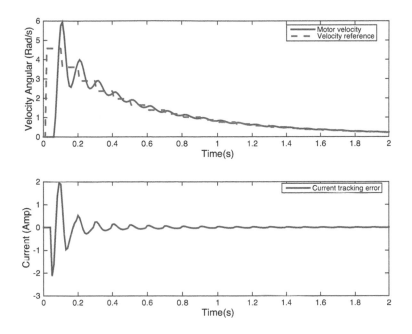

FIGURE 5.10: Trajectory tracking of the back left motor.

5.7 Conclusions

In this chapter a discrete-time decentralized neural inverse optimal control has been presented. The proposed controller is able to achieve the tracking of a system and is inverse optimal in the sense that, a posteriori, it minimizes a meaningful cost functional. The training of the neural network is performed on-line using an extended Kalman filter. A discrete-time application with an omnidirectional mobile robot illustrates the applicability of the proposed control techniques. The reference for the neural network controller is obtained through visual feedback. The simulations results show the effectiveness of the proposed approach.

6

Neural Network-Based Controller for Unmanned Aerial Vehicles

CONTENTS

6.1   Introduction

In recent years Unmanned Aerial Vehicles (UAVs) have demonstrated their capabilities and their potential to be used in several applications. The UAVs can move freely in a 3D space, which is a tremendous advantage over ground vehicles. Today they have many applications like mapping, inspection, surveying, aerial photography, agriculture, search and rescue robots, etc. Currently, there are many types of UAVs, they can be classified as fixed wing, single rotor, and multirotor. In this work, we focus in multirotor UAVs, and in particular we focus in quadrotors and hexarotors.

The quadrotor is an UAV with a center body with four arms, each of which has a motor with a propeller. These four propellers provide the thrust and lift of the vehicle. The quadrotor is a nonlinear underactuated system; it has 6 DOF, and only 4 control inputs (thrust, roll, pitch and yaw).

The hexarotor has some advantages over the quadrotor, such as, their increased load capacity, higher speed and safety, because of the fact that the two extra rotors, which allow the UAV to fly even if it loses one of the motors. The hexacopter is also a highly nonlinear and under actuated system.

One of the most common controllers in industry is the PID controller. The PID is used in many applications due to its simplicity. However the conventional PID controller is not effective when controlling large inertia systems [109, 141]. In contrast, an Artificial Neural Network (ANN) has many interesting properties like adaptability and learning capabilities. The ANN are used to control complex nonlinear systems [44]. Despite its limitations, a PID controller can cover most of its disadvantages when it is combined with an ANN [123].

In this chapter we present a neural network-based controller for multirotors. The advantage of this controller over conventional PID is that the neural network can tune the PID even if the multirotor changes its parameters during the flight and is not necessary to linearize the UAV model.

This chapter is organized as follows: In Section 6.2 the model of the quadrotor is introduced. In Section 6.3 the model of the hexarotor is presented. The design of the PID controller and weights adjustment are presented in Section 6.4. In Section 6.5 the visual servo control approach is introduced. Then the simulation results are shown in Section 6.6. The experimental results are presented in Section 6.7. Finally, the conclusions are given in Section 6.8.

6.2   Quadrotor Dynamic Modeling

A quadrotor with two pairs of propellers (1,3) and (2,4) in cross configuration and turning in opposite directions as shown in Fig. 6.1. The vertical motion of the quadrotor is given by a simultaneous increase or decrease of the speed of all the motors, Fig. 6.1.a. A roll rotation is generated by changing the speed of the propellers (2,4), Fig. 6.1.b. A pitch rotation is generated by changing the speed of the propellers (1,3), Fig. 6.1.c. The roll and the pitch rotations produce lateral motion. Finally, the yaw rotation is produced by increasing and decreasing one pair of the rotors, Fig. 6.1.d.

Let us assume that the center of mass and the body fixed frame origin coincide as in Fig. 6.2. The quadrotor orientation in space is given by a rotation matrix $R \in SO(3)$ from the body fixed frame to the inertial frame. The dynamics of a rigid body under external forces are expressed as

$$\begin{bmatrix} mI_{3\times3} & 0 \\ 0 & I \end{bmatrix} \begin{bmatrix} \dot{V} \\ \dot{\omega} \end{bmatrix} + \begin{bmatrix} \omega \times mV \\ \omega \times I\omega \end{bmatrix} = \begin{bmatrix} F \\ \tau \end{bmatrix} \tag{6.1}$$

where I is the inertia matrix; V the body linear speed vector and ω the angular speed.

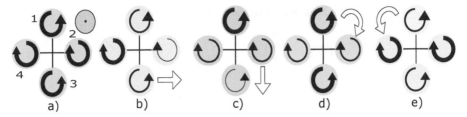

FIGURE 6.1: Quadrotor motion. The arrow width in the propeller is proportional to its rotation. White arrow describes the motion of the quadrotor.

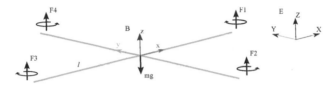

FIGURE 6.2: Quadrotor configuration. B represents the quadrotor fixed frame and E the inertial frame.

The quadrotor equations of motion can be expressed as

$$\dot{\zeta} = v$$
$$\dot{v} = -ge_3 + R_{e3}\left(\frac{b}{m}\sum\Omega_i^2\right)$$
$$\dot{R} = R\hat{\omega}$$
$$I\dot{\omega} = -\omega \times I\omega - \sum J_r\left(\omega \times e_3\right)\Omega_i + \tau_a$$

(6.2)

where ζ is the position vector, R the rotation matrix, $\hat{\omega}$ the skew symmetric matrix, Ω_i the rotor speed, I the body inertia, J_r the rotor inertia, b the thrust factor, d the drag factor, l distance from the body fixed frame origin to the rotor and τ_a the torque applied to the quadrotor and is expressed as

$$\tau_a = \begin{pmatrix} lb\left(\Omega_4^2 - \Omega_2^2\right) \\ lb\left(\Omega_3^2 - \Omega_1^2\right) \\ d\left(\Omega_2^2 + \Omega_4^2 - \Omega_1^2 - \Omega_3^2\right) \end{pmatrix}$$

(6.3)

The full quadrotor dynamic model is defined as

$$\ddot{x} = \left(\cos\left(\phi\right)sin\left(\theta\right)\sin\left(\psi\right) + \sin\left(\phi\right)\sin\left(\psi\right)\right)\frac{U_1}{m}$$
$$\ddot{y} = \left(\cos\left(\phi\right)sin\left(\theta\right)\sin\left(\psi\right) - \sin\left(\phi\right)\sin\left(\psi\right)\right)\frac{U_1}{m}$$
$$\ddot{z} = -g + \left(\cos\left(\phi\right)\cos\left(\theta\right)\right)\frac{U_1}{m}$$
$$\ddot{\phi} = \dot{\theta}\dot{\psi}\left(\frac{I_y - I_z}{I_x}\right) - \frac{J_r}{I_x}\dot{\theta}\Omega + \frac{l}{I_x}U_2$$
$$\ddot{\theta} = \dot{\phi}\dot{\psi}\left(\frac{I_z - I_x}{I_y}\right) + \frac{J_r}{I_y}\dot{\phi}\Omega + \frac{l}{I_y}U_3$$
$$\ddot{\psi} = \dot{\phi}\dot{\theta}\left(\frac{I_x - I_y}{I_z}\right) + \frac{U_4}{I_z}$$

(6.4)

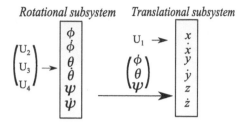

FIGURE 6.3: U_2, U_3 and U_4 are inputs for the rotational subsystem; U_1, roll, pitch and yaw are inputs for the following translation subsystem.

with U_i as the system's inputs and Ω the changing attitude angle which is part of the gyroscopic effects induced by the propellers. Gyroscopic effects provide a more accurate model; nevertheless, they have insignificant roles in the overall attitude of the quadcopter [122]. The inputs of the system are defined as

$$
\begin{aligned}
U_1 &= b\left(\Omega_1^2 + \Omega_2^2 + \Omega_3^2 + \Omega_4^2\right) \\
U_2 &= b\left(\Omega_4^2 - \Omega_2^2\right) \\
U_3 &= b\left(\Omega_3^2 - \Omega_1^2\right) \\
U_4 &= b\left(\Omega_2^2 + \Omega_4^2 - \Omega_1^2 - \Omega_3^2\right) \\
\Omega &= \Omega_2 + \Omega_4 - \Omega_1 - \Omega_3
\end{aligned}
\tag{6.5}
$$

The quadrotor is a rotating rigid body with six degrees of freedom and rotational and translational dynamics. Inputs U_i and their relation with both subsystems are shown in Fig. 6.3.

6.3   Hexarotor Dynamic Modeling

The hexarotor consists of six arms connected symmetrically to the central hub, each of which has a motor and a propeller. Each propeller produces an upward thrust and since they are located outside the center of gravity, differential thrust is used to rotate the hexarotor. In addition, the rotation of the propellers also produces a torque in the opposite direction of the motor rotation, therefore, there must be two groups of rotors spinning in opposite directions for the purpose of making this reaction torque equal to zero.

The pose of an hexarotor is given by its position $\zeta = [x, y, z]^T$ and its orientation $\eta = [\phi, \theta, \psi]^T$ in the three Euler angles roll, pitch and yaw respectively. For the sake of simplicity, $sin(\cdot)$ and $cos(\cdot)$ will be abbreviated as $s\cdot$ and $c\cdot$. The transformation from world frame O to body frame (Fig. 6.4) is given by

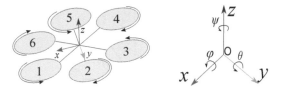

FIGURE 6.4: Structure of hexarotor and coordinate frames.

$$
\begin{bmatrix} x_B \\ y_B \\ z_B \end{bmatrix} = \begin{bmatrix} c\theta c\psi & c\theta s\psi & -s\theta \\ s\phi s\theta c\psi - c\phi s\psi & s\phi s\theta s\psi + c\phi c\psi & s\phi c\theta \\ c\phi s\theta c\psi + s\phi s\psi & c\phi s\theta s\psi - s\phi c\psi & c\phi c\theta \end{bmatrix} \begin{bmatrix} x_W \\ y_W \\ z_W \end{bmatrix} \tag{6.6}
$$

The dynamic model of the robot expressed in the body frame in Newton-Euler formalism is obtained as in [89].

$$
\begin{bmatrix} mI_{3\times3} & 0 \\ 0 & I \end{bmatrix} \begin{bmatrix} \dot{V} \\ \dot{\omega} \end{bmatrix} + \begin{bmatrix} \omega \times mV \\ \omega \times I\omega \end{bmatrix} = \begin{bmatrix} F \\ \tau \end{bmatrix} \tag{6.7}
$$

where I is the 3×3 inertia matrix; V the linear speed vector and ω the body angular speed. The equations of motion for the helicopter of Fig. 6.4 can be written as in [18]

$$
\begin{aligned}
\dot{\zeta} &= v \\
\dot{v} &= -ge_3 + R\left(\frac{b}{m}\sum \Omega_i^2\right) \\
\dot{R} &= R\hat{\omega} \\
I\dot{\omega} &= -\omega \times I\omega - \sum J_r\left(\omega \times e_3\right)\Omega_i + \tau_a
\end{aligned} \tag{6.8}
$$

where ζ is the position vector, R is the rotation matrix from the body frame to the world frame, $\hat{\omega}$ represents the skew symmetric matrix, Ω is the rotor speed, I is the body inertia, J_r is the rotor inertia, b is the thrust factor and τ is the torque applied to the body frame due to the rotors. Since we are dealing with an hexarotor, this torque vector differs from the well known quadrotor torque vector and if we are working with a structure like the one in Fig. 6.5 it can be written as

$$
\tau_a = \begin{pmatrix} bl\left(-\Omega_2^2 + \Omega_5^2 + \frac{1}{2}\left(-\Omega_1^2 - \Omega_3^2 + \Omega_4^2 + \Omega_6^2\right)\right) \\ bl\frac{\sqrt{3}}{2}\left(-\Omega_1^2 + \Omega_3^2 + \Omega_4^2 - \Omega_6^2\right) \\ d\left(-\Omega_1^2 + \Omega_2^2 - \Omega_3^2 + \Omega_4^2 - \Omega_5^2 + \Omega_6^2\right) \end{pmatrix} \tag{6.9}
$$

where l is the distance from the center of gravity of the robot to the rotor and

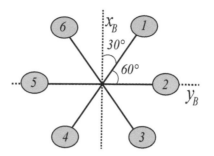

FIGURE 6.5: Geometry of hexarotor.

b is the thrust factor. The full dynamic model is

$$\ddot{x} = (\cos\phi sin\theta cos\psi + \sin\phi sin\psi)\frac{U_1}{m}$$

$$\ddot{y} = (\cos\phi sin\theta \sin\psi - \sin\phi\cos\psi)\frac{U_1}{m}$$

$$\ddot{z} = -g + (\cos\phi\cos\theta)\frac{U_1}{m}$$

$$\ddot{\phi} = \dot{\theta}\dot{\psi}\left(\frac{I_y - I_z}{I_x}\right) - \frac{J_r}{I_x}\dot{\theta}\Omega + \frac{l}{I_x}U_2 \qquad (6.10)$$

$$\ddot{\theta} = \dot{\phi}\dot{\psi}\left(\frac{I_z - I_x}{I_y}\right) - \frac{J_r}{I_y}\dot{\phi}\Omega + \frac{l}{I_y}U_3$$

$$\ddot{\psi} = \dot{\phi}\dot{\theta}\left(\frac{I_x - I_y}{I_z}\right) + \frac{l}{I_z}U_4$$

where U_1, U_2, U_3, U_4 and Ω represent the system inputs and in the case of the hexarotor are obtained as follows

$$U_1 = b\left(\Omega_1^2 + \Omega_2^2 + \Omega_3^2 + \Omega_4^2 + \Omega_5^2 + \Omega_6^2\right)$$

$$U_2 = bl\left(-\frac{\Omega_1^2}{2} - \Omega_2^2 - \frac{\Omega_3^2}{2} + \frac{\Omega_4^2}{2} + \Omega_5^2 + \frac{\Omega_6^2}{2}\right)$$

$$U_3 = bl\left(-\frac{\sqrt{3}}{2}\Omega_1^2 + \frac{\sqrt{3}}{2}\Omega_3^2 + \frac{\sqrt{3}}{2}\Omega_4^2 - \frac{\sqrt{3}}{2}\Omega_6^2\right) \qquad (6.11)$$

$$U_4 = d\left(-\Omega_1^2 + \Omega_2^2 - \Omega_3^2 + \Omega_4^2 - \Omega_5^2 + \Omega_6^2\right)$$

where d is the drag factor.

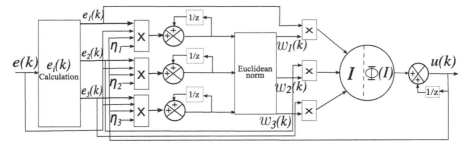

FIGURE 6.6: PID-ANN topology. There is one module PID-ANN for every degree of freedom.

6.4 Neural Network-Based PID

The conventional PID with unitary feedback is described in [96], where the control law is defined as

$$U(z) = \left[K_P + \frac{K_I}{(1 - z^{-1})} + K_D \left(1 - z^{-1}\right) \right] E(z) \tag{6.12}$$

where $E(z)$ is the error calculated as the difference between the reference signal and the system output $R(z) - Y(z)$. The terms K_P, K_I, and K_D are the proportional, integral and derivative gains respectively. Although conventional PID is widely used, due to its simplicity and performance, it is not always a good technique when controlling highly nonlinear systems, such as multirotors.

To handle the nonlinearities of the multirotors an ANN-based PID controller is used. The ANN not only can deal with the nonlinearities of the system, but also, it can adjust the gains of the PID controller. These capabilites can also handle the uncertainties of the model. The topology of the PID-ANN used is shown in Fig. 6.6.

From Fig. 6.6, the $e_i(k)$ vector represents the derivative of the error, the proportional error and the sum of the errors. They are defined as follows:

$$
\begin{aligned}
e_1(k) &= e(k) \\
e_2(k) &= e(k) - 2e(k-1) + e(k-2) \\
e_3(k) &= e(k) - e(k-1)
\end{aligned}
\tag{6.13}
$$

Accordingly, the control law of the conventional PID can be rewritten as

$$u(k) = u(k-1) + K_P e_1(k) + K_I e_2(k) + K_D e_3(k) \tag{6.14}$$

The Neuron input is defined as

$$I = \sum_{i=1}^{3} e_i(k) w_i(k) \tag{6.15}$$

where vector $w_i(k)$ represents the weights of the network which are incremented by

$$\Delta w_i = \eta_i e_i(k) e(k) u(k) \tag{6.16}$$

with a learning factor η. The new value of $w_i(k)$ will be

$$w_i'(k) = w_i(k-1) + \Delta w_i(k) \tag{6.17}$$

The Euclidean norm will be used to limit the values of $w_i(k)$ as

$$w_i(k) = \frac{w_i'(k)}{\left\| \sum_{i=1}^{3} w_i'(k) \right\|} \tag{6.18}$$

The activation function of the neuron is the hyperbolic tangent, therefore the output will be

$$\Phi(I) = A\frac{1 - e^{-Ib}}{1 + e^{-Ib}} \tag{6.19}$$

where A is a gain factor to escalate the maximum value of the activation function which is between $[-1, 1]$ and b is a scalar to avoid saturation of the neuron. The control law of the ANN-based PID is expressed as follows

$$u(k) = u(k-1) + \Phi(I) \tag{6.20}$$

and there is one PID-ANN module for every U_i to control in (6.11).

6.5 Visual Servo Control

In this work an Image-Based Visual Servo (IBVS) control approach is used. The camera is mounted on the robot and the movement of the hexarotor induces camera motion [26].

The visual serving purpose is the minimization of the error

$$e(t) = s(m(t), a) - s^* \tag{6.21}$$

where $m(t)$ is a vector of 2D points coordinates in image plane and a is a set of known parameters of the camera (e.g. camera intrinsic parameters). Vector s^* contains the desired values. Since the error $e(t)$ is defined on the image space and the robot moves in the 3D space, it is necessary to relate changes in the image features with the hexarotor displacement. The image Jacobian [138] (also known as interaction matrix) captures the relation between features and robot velocities as shown

$$\dot{s} = L_s v_c \tag{6.22}$$

where $\dot{s}$ is the variation of the features position and $\mathrm{v_c} = (v_c, \dot{\omega}_c)$ denotes the camera translational $\dot{v}_c$ and rotational $\dot{\omega}_c$ velocities. Considering $\mathrm{v_c}$ as the control input, we can try to ensure an exponential decrease of the error with

$$\mathrm{v}_c = -\lambda \mathrm{L_s}^+ e \tag{6.23}$$

where λ is a positive constant, $\mathrm{L}_s \in \mathbb{R}^{6 \times k}$ is the pseudo-inverse of L_s, k is the number of features and e the feature error.

To calculate L consider a 3D point X with coordinates (X, Y, Z) in the camera frame, the projected point in the image plane x with coordinates (x, y) is defined as

$$\begin{aligned} x &= X/Z = (u - c_u)/f\alpha \\ y &= Y/Z = (v - c_v)/f \end{aligned} \tag{6.24}$$

where (u, v) are the coordinates of the point in the image space expressed in pixel units, (c_u, c_v) are the coordinates of the principal point, α is the ratio of pixel dimensions and f the focal length.

The feature velocity is defined as

$$\dot{\mathrm{x}} - \mathrm{L}_x \mathrm{v}_c \tag{6.25}$$

where

$$\mathrm{L}_x = \begin{bmatrix} -\frac{1}{Z} & 0 & \frac{x}{Z} & xy & -(1+x^2) & y \\ 0 & -\frac{1}{Z} & \frac{y}{Z} & 1+y^2 & -xy & -x \end{bmatrix} \tag{6.26}$$

where Z is the depth of the feature. In our case we use a RGB-D sensor and this distance is known. To control the 6 DOF, at least three points are necessary [26], in that particular case, we would have three interaction matrices L_{x_1}, L_{x_2}, L_{x_3}, one for each feature, and the complete interaction matrix is now $L_x = \begin{bmatrix} L_{x_1} L_{x_2} L_{x_3} \end{bmatrix}^T$. When using three points, there are some configurations for which L_x is singular and four global minima [87]. More precisely, there are four poses for the camera such that $\dot{s} = 0$, these four poses are impossible to differentiate [42]. With this in mind, it is usual to consider more points [26].

6.5.1 Control of hexarotor

The hexarotor has four control inputs U_i; U_1 represents the translation in the z axis, U_2 represents the roll torque, U_3 represents the pitch and U_4 represents the yaw torque. The visual algorithm acts as a proportional controller where λ in (6.23) works as a proportional gain. When combined with the ANN based PID, we can adapt not only this proportional gain but also the derivative and the integral gains. Since the system is underactuated, we can use the translational velocities $[\dot{x}, \dot{y}]$ computed by IBVS as input roll and pitch torques and the error will be reduced. This is shown in Fig. 6.7.

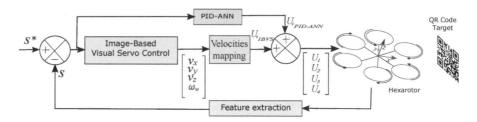

FIGURE 6.7: Block diagram of our IBVS algorithm combined with ANN-PID.

6.6   Simulation Results

In this section we present the simulations results with the quadrotor and with the hexarotor.

6.6.1   Quadrotor simulation results

The simulation experiments were carried out using MATLAB with the Robotics Toolbox [29]. There is one PIDNN subsystem for each of the controlling variables: position in x and y, altitude in z and the yaw angle ψ. In the first experiment, only altitude was controlled at a constant value (5m), and then the mass of the quadrotor was increased by 100% at $t = 10s$ and again at $t = 15s$. As can be seen in Fig. 6.8, the ANN adapts its weights (PID gains) to keep the reference.

In the second experiment, we add a circular trajectory on the xy plane and the mass of the quadrotor is changed (increased by 20%). The plots in Fig. 6.9 show the results of this experiment. As shown, the conventional PID controller is unable to follow the reference when the model is changed.

In the last experiment, yaw angle is added to the xy displacement, the results are shown in Fig. 6.10. As in the last simulation, the mass in the quadrotor is increased. It can be seen that the quadrotor trajectory is completely lost when the yaw angle is added and controlled with conventional PID. Conversely, PIDNN still follows the reference even when the mass is changed and a yaw angle is requested.

6.6.2   Hexarotor simulation results

The simulation experiments were implemented on MATLAB using the Robotics Toolbox [29]. For the visual servo algorithm, four points were used. In the first experiment, the robot starts on the ground and has to reach a certain position given by these 2D points. In the simulation at the second 10, the mass of the robot has been increased 50%. It can be seen that the

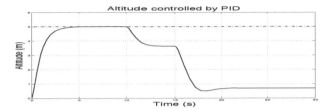

(a) Height controlled by conventional PID

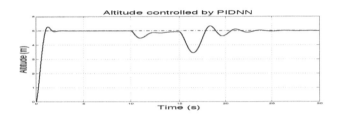

(b) Height controlled by PIDNN while quadrotor mass changes

FIGURE 6.8: Reference remains constant at $5m$. Dashed lines represent the reference and solid lines represent the system output.

conventional PID controller is unable to keep position. The results are shown in Fig. 6.11.

In the second experiment, the mass of the system remains constant but the moment of inertia I_x was increased from $I_x = 0.0820$ to $I_x = 0.550$. Figure 6.12 shows the results of using conventional PID when the moment of inertia is increased. The results show that the control input is excessively high to the robot (Fig. 6.12(b)), making the system unable to follow the reference (Fig. 6.12(c)).

In the following experiments, the PID-ANN is now controlling the system under the same conditions. The mass has been incremented at second 10. It can be seen in Fig. 6.13 that the controller can be adapted to this mass increment and keep the reference.

Finally, we show, in Fig. 6.14 the results of changing the moment of inertia I_x while mass remains constant. In contrast with conventional PID, the PID-ANN is able to keep its position.

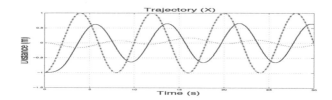

(a) *X* position controlled with PID and PIDNN

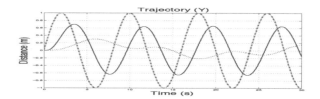

(b) *Y* position controlled with PID and PIDNN

FIGURE 6.9: Both controllers have the same initial gains and they were tuned for a quadrotor with different mass. Dashed lines represent the reference, solid lines represent the system output controlled by PIDNN and dotted lines represent the system output controlled by conventional PID.

6.7 Experimental Results

In this section we present the experimental results with the quadrotor and with the hexarotor. The quadrotor used in the experiments was a Quanser Qball, and the hexarotors was an Astec Firefly.

6.7.1 Quadrotor experimental results

Due to the obtained simulation results, the PIDNN controller was tested on the Quanser Qball X4, which is an indoor vehicle suitable for research applications. The Qball X4 is propelled by four brushless motors and equipped with a 3-axis accelerometer, 3-axis gyroscope and a sonar height sensor. For translational measures, an Optitrack Camera System was used, it consists of six synchronized infrared cameras tracking the quadrotor position and velocity in x, y and z. Both the quadrotor and cameras system were controlled by a ground station (Fig. 6.15). Actual configuration can be seen in Fig. 6.16.

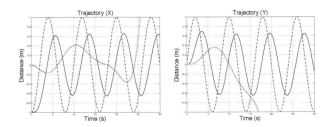

(a) X position controlled simultaneously with PID and
PIDNN

(b) Y position controlled simultaneously with PID and
PIDNN

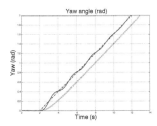

(c) Yaw angle controlled simultaneously with PID and
PIDNN

FIGURE 6.10: Both controllers have the same initial gains and they were
tuned for a quadrotor with different mass. Dashed lines represent the reference,
solid lines represent the system output controlled by PIDNN and dotted lines
represent the system output controlled by conventional PID.

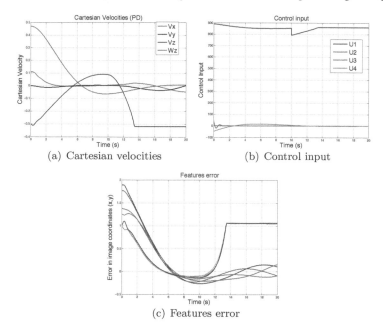

(a) Cartesian velocities   (b) Control input

(c) Features error

FIGURE 6.11: Simulation using conventional PID. At 10 seconds the mass of the system is incremented 50% and $\lambda = 0.3$.

The actual relative positions between cameras and Qball X4 can be seen with the Optitrack software (Motive) and is shown in Fig. 6.17.

In the experiments, altitude, position in x, position in y and yaw angle were controlled separately. Quanser provides a conventional PID controller which was tested and is presented in Fig. 6.18. Although Qball can follow the reference in altitude and yaw when it is controlled with a conventional PID controller, it presents more than 30% overshoot when following a trajectory in x and y position making the system slower and unable to follow the reference. Even though a change in the PID gains could lead to a lower overshoot or faster response, for the second experiment the PIDNN will control the Qball with the same initial gains and the ANN will adapt them.

As shown in Fig. 6.19, no significant differences in height were found and even when the response in yaw angle is slower than the PID, it can follow the reference. The main contribution of the neural network can be seen in position x and y where the controller gains have been adapted and overshoot has been eliminated. The system is faster when controlled by PIDNN and it makes the Qball able to correctly follow the reference without concern for the correct PID tuning.

Now that PIDNN has shown faster response, overshoot has been eliminated in position and reduced in height, and for a third experiment, the quadrotor followed a path in x while keeping at $y = 0$ (Fig. 6.20 (a) and (b)) which is consistent with the plots in Fig. 6.20 (c) and (d) showing the roll and pitch angle which are responsible of the translational displacement of the Qball. Desired roll and pitch angles are computed with inverse kinematics [51]. At the same time, the system changes its height (Fig. 6.20 (e)) and rotates in z axis (Fig. 6.20 (f)); as in the last experiment where each state is controlled separately, a 10% overshoot is present in height and it follows the yaw angle reference.

To compare the behavior of the conventional PID and the ANN based PID we use the Root Mean Squared Error (RMSE) and Average Absolute Deviation (AAV) of error, the results shown in Table 6.1.

TABLE 6.1: Controllers comparison.

		RMSE				AAD		
	x	y	Height	Yaw	x	y	Height	Yaw
PID	0.8885	0.3111	0.1697	0.3636	0.3473	0.2930	0.8858	0.4357
PIDNN	0.2427	0.2396	0.1593	0.2237	0.3314	0.2562	0.8151	0.2483

6.7.2 Hexarotor experimental results

The hexarotor used in the experiments was an Asctec Firefly. The actual configuration of the experiment is shown in Fig. 6.21. The vision sensor used is an Intel RealSense R200 camera with RGB-D sensor with a range from 0.4m to 2.8m. It is important to note that vision information is highly noisy and presents high computational cost even when working with low resolution images (in this case 640 × 480). The more time the algorithm uses the image capture and processing phase, the more error will exist between what the robot sees and the actual position; coordination between the vision sensors, neural network and model system working at different processing stages and its communication at their respective architectures is crucial to achieve real-time implementation. A QR-code was used as the tracking pattern; this pattern was chosen because of its robustness to rotations and illumination changes. The algorithms were implemented on the onboard computer of the hexarotor.

In the first experiment, the moment of inertia and mass of the system changed due to the addition of the new sensor. The original PID with the factory parameters was compared with the proposed approach. Figure 6.22 shows results when the pattern is fixed at a certain position and the hexarotor is at hover position. As it can be seen in Fig. 6.22(b), a conventional controller cannot achieve system stabilization at a fixed position when the model changes. Table 6.2 shows the Root Mean Square Error (RMSE) and the Average Absolute Deviation (AAD) in pixel units. The pair (x_i, y_i) are the location of the feature ($i = 1, 2, 3, 4$) in image coordinates.

In Fig. 6.22(b), the solid increasing lines represent the x position of the 4 features in image coordinates (pixel units). As can be seen, if a conventional PID is not correctly tuned for this specific system its position diverges. On the other hand, when the system is controlled by the PID-ANN its position does not diverge (Fig. 6.22(d)) even when the controller has not been previously tuned.

TABLE 6.2: Controller comparison.

	x_1	y_1	x_2	y_2	x_3	y_3	x_4	y_4
Root Mean Square Error								
PID	1741.9	393.1	1712.0	339.1	1677.5	329.2	1723.9	380.1
PIDNN	212.416	66.549	191.1447	57.4298	824.085	59.7654	861.8207	62.6225
Average Absolute Deviation								
	x_1	y_1	x_2	y_2	x_3	y_3	x_4	y_4
PID	55.2177	9.6437	54.0636	10.5457	52.4827	10.704	53.069	9.6577
PIDNN	43.2281	7.4259	42.5686	11.3874	46.6832	11.484	47.5576	7.5503

Once the ANN-PID demonstrated its effectiveness over the PID controller, the experiment was repeated but now the QR pattern was in motion. As shown in Fig. 6.23 the hexarotor did not lose sight of the objective.

6.8 Conclusions

In this work a neural network-based PID controller for multirotors was presented. As shown in the results, the proposed PIDNN can control a multirotor in real time. Although conventional PID can control z axis and yaw, it presents problems when translating in x and y even when every state was tested separately. In contrast, the multirotors can follow a trajectory when they are controlled by the PIDNN. Another advantage is that the ANN can deal with the nonlinearity of the system. In addition, it is not necessary to tune the controller; the ANN can set the PID gains.

In the hexarotor experiments a neural network-based PID controller with visual feedback was presented. The hexarotor was equipped with a RGB-D sensor that allows it to estimate the feature error; this error has been used to compute the camera velocities. The proposed approach is able to deal with delays due to image processing, system uncertainties and small changes in the model since the ANN is continuously adapting the PID gains. In contrast to conventional PID controllers, where it is mandatory to tune it according to a specific system, the ANN can deal with nonlinearities and changes in the system.

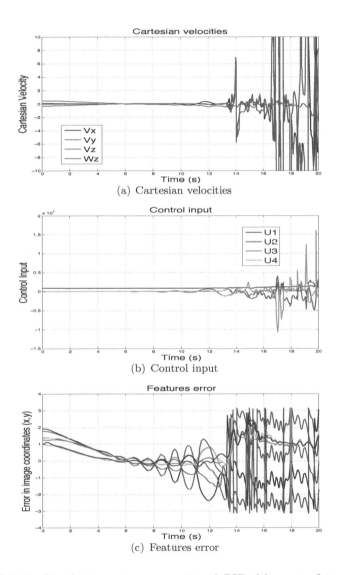

FIGURE 6.12: Simulation using conventional PID. Moment of inertia is increased. Mass remains constant. It can be seen that the robot is not stable when I_x changes.

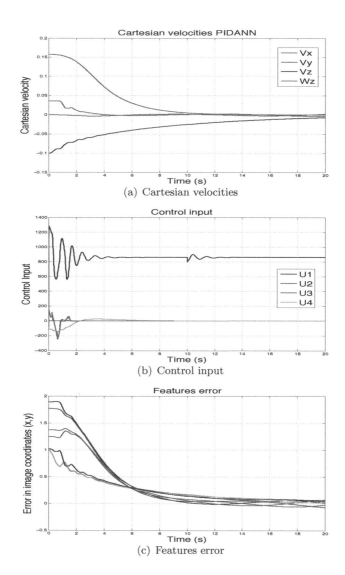

(a) Cartesian velocities

(b) Control input

(c) Features error

FIGURE 6.13: Simulation using conventional PID-ANN. At 10 seconds the mass of the system is incremented 50% and $\lambda = 0.3$. System remains at desired position.

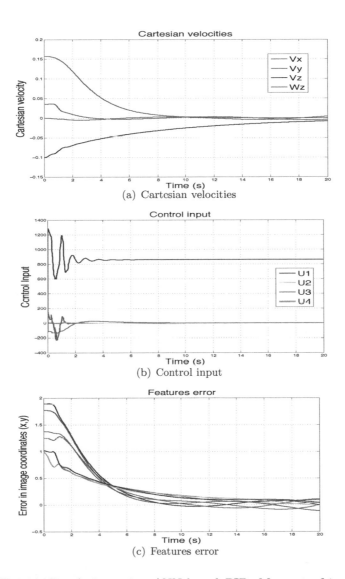

FIGURE 6.14: Simulation using ANN-based PID. Moment of inertia is increased. Mass remains constant. It can be seen that the robot remains stable at desired position when I_x changes.

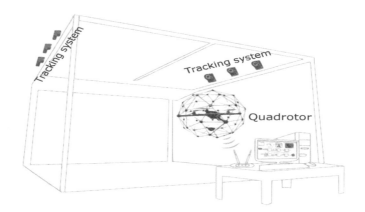

FIGURE 6.15: Experiment configuration.

FIGURE 6.16: Actual laboratory configuration.

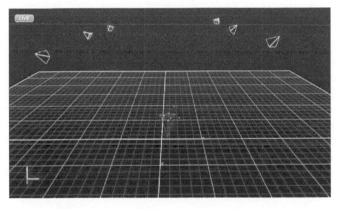

FIGURE 6.17: Experiment configuration with actual relative position from Optitrack software. The rigid body in the origin represents the Qball X4. The six floating squares represent the cameras.

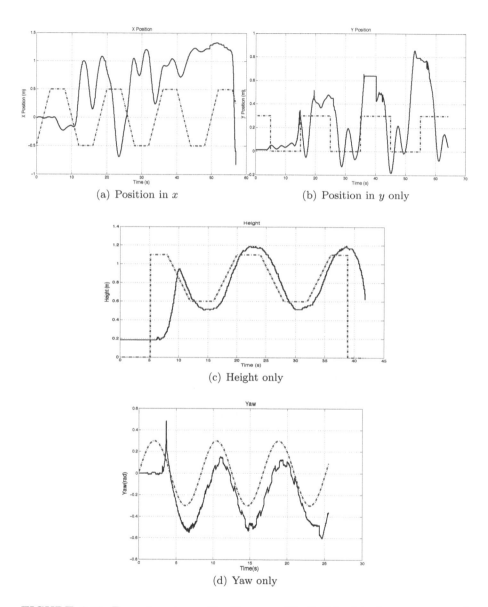

(a) Position in x

(b) Position in y only

(c) Height only

(d) Yaw only

FIGURE 6.18: Experiments on the Quanser Qball X4 with the conventional PID. Each degree of freedom was tested separately. Dashed lines represent the reference and solid lines represent the system output.

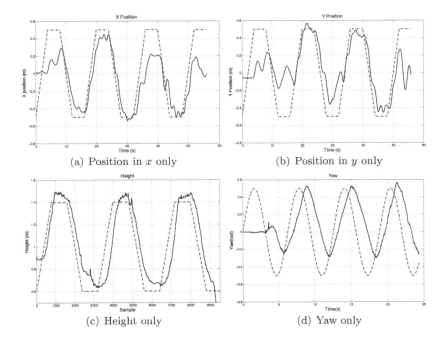

(a) Position in x only

(b) Position in y only

(c) Height only

(d) Yaw only

FIGURE 6.19: Experiments on the Quanser Qball X4 controlled with PIDNN. Dashed lines represent the reference and solid lines represent system output.

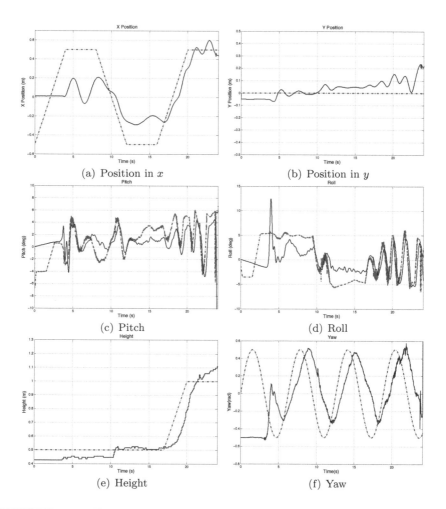

FIGURE 6.20: Experiments on the Quanser Qball X4. Degrees of freedom combined and controlled with PIDNN. Dashed lines represent the reference and solid lines represent system output.

FIGURE 6.21: Actual experiment configuration. The corners of the QR code represent the 3D features.

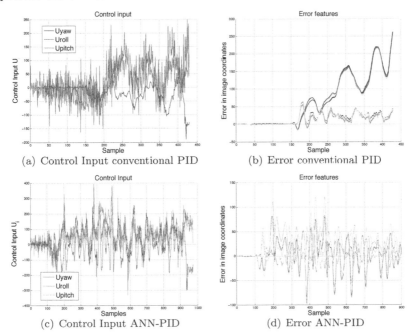

(a) Control Input conventional PID

(b) Error conventional PID

(c) Control Input ANN-PID

(d) Error ANN-PID

FIGURE 6.22: Experimental results when mass and moment of inertia changed. The pattern (3D features) remains at the same position during the test.

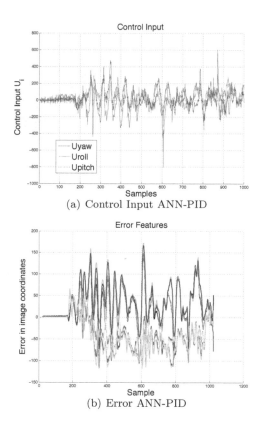

(a) Control Input ANN-PID

(b) Error ANN-PID

FIGURE 6.23: Hexarotor experimental results at hover position when mass and moment of inertia changed. The pattern (3D features) is moving during the test.

Bibliography

[1] S. Akhavan and M. Jamshidi. ANN-based sliding mode control for non-holonomic mobile robots. In *Control Applications, 2000. Proceedings of the 2000 IEEE International Conference on*, pages 664–667, 2000.

[2] A. Al-Tamimi, F.L. Lewis, and M. Abu-Khalaf. Discrete-time non-linear hjb solution using approximate dynamic programming: Convergence proof. *Systems, Man, and Cybernetics, Part B: Cybernetics, IEEE Transactions on*, 38(4):943–949, Aug 2008.

[3] Alma Y Alanis, Nancy Arana-Daniel, Carlos Lopez-Franco, and Edgar Guevara-Reyes. Integration of an inverse optimal neural controller with reinforced-slam for path panning and mapping in dynamic environments. *Computación y Sistemas*, 19(3):445–456, 2015.

[4] Alma Y. Alanis, Michel Lopez-Franco, Nancy Arana-Daniel, and Carlos Lopez-Franco. Discrete-time neural control for electrically driven non-holonomic mobile robots. *International Journal of Adaptive Control and Signal Processing*, 26(7):630–644, 2012.

[5] Alma Y. Alanis, Jorge D. Rios, Nancy Arana Daniel, and Carlos Lopez Franco. Neural identifier for unknown discrete-time nonlinear delayed systems. *Neural Computing and Applications*, 2015.

[6] Alma Y. Alanis, Jorge D. Rios, Jorge Rivera, Nancy Arana-Daniel, and Carlos Lopez-Franco. Real-time discrete neural control applied to a Linear Induction Motor. *Neurocomputing*, 164, 2015.

[7] J. Andrade-Cetto and Alberto Sanfeliu. Temporal landmark validation in cml. In *Robotics and Automation, 2003. Proceedings. ICRA '03. IEEE International Conference on*, volume 2, pages 1576–1581 vol.2, 2003.

[8] M. Elena Antonio-Toledo, E. N. Sanchez, and A. G. Loukianov. Real-time implementation of a neural block control using sliding modes for induction motors. In *2014 World Automation Congress (WAC)*, pages 502–507, 2014.

[9] N. Arana-Daniel, R. Rosales-Ochoa, and C. Lopez-Franco. Reinforced-slam for path planing and mapping in dynamic environments. In *Electrical Engineering Computing Science and Automatic Control (CCE), 2011 8th International Conference on*, pages 1–6, 2011.

[10] Nancy Arana-Daniel, Carlos Villaseñor, Carlos López-Franco, and Alma Y Alanís. Bio-inspired aging model-particle swarm optimization and geometric algebra for structure from motion. In *Iberoamerican Congress on Pattern Recognition*, pages 762–769. Springer, 2014.

[11] Vladimir Banarer, Christian Perwass, and Gerald Sommer. The hypersphere neuron. In *ESANN*, pages 469–474, 2003.

[12] T. Basar and G. J. Olsder. *Dynamic Noncooperative Game Theory*. Academic Press, New York, New York, USA, 2nd edition, 1995.

[13] Herbert Bay, Andreas Ess, Tinne Tuytelaars, and Luc Van Gool. Speeded-up robust features (surf). *Comput. Vis. Image Underst.*, 110(3):346–359, June 2008.

[14] Eduardo Bayro-Corrochano and Miguel Bernal-Marin. Generalized hough transform and conformal geometric algebra to detect lines and planes for building 3d maps and robot navigation. In *Intelligent Robots and Systems (IROS), 2010 IEEE/RSJ International Conference on*, pages 810–815. IEEE, 2010.

[15] Eduardo Bayro-Corrochano and Sven Buchholz. Geometric neural networks. In *Algebraic Frames for the Perception-Action Cycle*, pages 379–394. Springer, 1997.

[16] Eduardo Bayro-Corrochano, Leo Reyes-Lozano, and Julio Zamora-Esquivel. Conformal geometric algebra for robotic vision. *Journal of Mathematical Imaging and Vision*, 24(1):55–81, 2006.

[17] Eduardo Jose Bayro-Corrochano and Nancy Arana-Daniel. Clifford support vector machines for classification, regression, and recurrence. *IEEE Transactions on Neural Networks*, 21(11):1731–1747, 2010.

[18] Samir Bouabdallah, Pierpaolo Murrieri, and Roland Siegwart. Design and control of an indoor micro quadrotor. In *Robotics and Automation, 2004. Proceedings. ICRA'04. 2004 IEEE International Conference on*, volume 5, pages 4393–4398. IEEE, 2004.

[19] P. Bouboulis, S. Theodoridis, C. Mavroforakis, and L. Evaggelatou-Dalla. Complex support vector machines for regression and quaternary classification. *IEEE Transactions on Neural Networks and Learning Systems*, 26(6):1260–1274, June 2015.

[20] Bernard Brogliato, Rogelio Lozano, Bernhard Maschke, and Olav Egeland. *Dissipative Systems Analysis and Control : Theory and Applications*. Communications and Control Engineering. Springer, London, 2007.

[21] Sven Buchholz and Gerald Sommer. Introduction to neural computation in clifford algebra. In *Geometric computing with Clifford algebras*, pages 291–314. Springer, 2001.

[22] R. J. Campbell and Y P. J. Flynn. A survey of free-form object representation and recognition techniques. *Computer Vision and Image Understanding*, 81:166–210, 2001.

[23] U. M. Ch, Y.S.K. Babu, and K. Amaresh. *Sliding Mode Speed Control of a DC Motor*. Communication Systems and Network Technologies (CSNT), Katra, Jammu, India, 2011.

[24] F. Chaumette and S. Hutchinson. Visual servo control. i. basic approaches. *Robotics Automation Magazine, IEEE*, 13(4):82–90, Dec 2006.

[25] F. Chaumette and S. Hutchinson. Visual servo control, part ii: Advanced approaches. *IEEE Robotics and Automation Magazine*, 14(1):109–118, March 2007.

[26] François Chaumette and Seth Hutchinson. Visual servo control. i. basic approaches. *IEEE Robotics & Automation Magazine*, 13(4):82–90, 2006.

[27] Z. Chen, B. Yao, and Q. Wang. μ-synthesis-based adaptive robust control of linear motor driven stages with high-frequency dynamics: A case study. *IEEE/ASME Transactions on Mechatronics*, 20(3):1482–1490, June 2015.

[28] Zheng Chen, Ya-Jun Pan, and Jason Gu. Integrated adaptive robust control for multilateral teleoperation systems under arbitrary time delays. *International Journal of Robust and Nonlinear Control*, 26(12):2708–2728, 2016. rnc.3472.

[29] Peter I. Corke. *Robotics, Vision & Control: Fundamental Algorithms in MATLAB*. Springer, 2011.

[30] Corinna Cortes and Vladimir Vapnik. Support-vector networks. *Machine Learning*, 20(3):273–297, 1995.

[31] T. Das and I. N. Kar. Design and implementation of an adaptive fuzzy logic-based controller for wheeled mobile robots. *IEEE Transactions on Control Systems Technology*, 14(3):501–510, 2006.

[32] K.D. Do, Z.P. Jiang, and J. Pan. Simultaneous tracking and stabilization of mobile robots: an adaptive approach. *Automatic Control, IEEE Transactions on*, 49(7):1147 – 1151, july 2004.

[33] Khac Duc Do, Jiang Zhong-Ping, and Jie Pan. Simultaneous tracking and stabilization of mobile robots: an adaptive approach. *Automatic Control, IEEE Transactions on*, 49(7):1147–1151, 2004.

[34] Ioan Doroftei, Victor Grosu, and Veaceslav Spinu. *Omnidirectional mobile robot–design and implementation.* INTECH Open Access Publisher, 2007.

[35] H. Durrant-Whyte and Tim Bailey. Simultaneous localization and mapping: part i. *Robotics Automation Magazine, IEEE*, 13(2):99–110, 2006.

[36] Hugh Durrant-Whyte and Tim Bailey. Simultaneous localization and mapping (slam), part ii: State of the art. *IEEE Robotics & Automation Magazine*, 2, 2006.

[37] B. Espiau, F. Chaumette, and P. Rives. A new approach to visual servoing in robotics. *Robotics and Automation, IEEE Transactions on*, 8(3):313–326, June 1992.

[38] L. A. Feldkamp, D. V. Prokhorov, and T. M. Feldkamp. Simple and conditioned adaptive behavior from Kalman. *Neural Networks*, 16:683–689, 2003.

[39] R. A. Felix. *Variable Structure Neural Control.* Ph.D thesis, Cinvestav, Unidad Guadalajara, Guadalajara, Jalisco, Mexico, 2003.

[40] Ramon A Felix, Edgar N Sanchez, and Guanrong Chen. Reproducing chaos by variable structure recurrent neural networks. *IEEE transactions on neural networks*, 15(6):1450–1457, 2004.

[41] R. Fierro and F. L. Lewis. Control of a nonholonomic mobile robot using neural networks. *IEEE Transactions on Neural Networks*, 9(4):589–600, 1998.

[42] Martin A Fischler and Robert C Bolles. Random sample consensus: a paradigm for model fitting with applications to image analysis and automated cartography. *Communications of the ACM*, 24(6):381–395, 1981.

[43] King San Fu, Rafael C. Gonzalez, and C. S. George Lee. *Robotics: control, sensing, vision, and intelligence.* McGraw-Hill,, New York, 1987.

[44] Shuzhi Sam Ge, Jin Zhang, and Tong Heng Lee. Adaptive neural network control for a class of mimo nonlinear systems with disturbances in discrete-time. *IEEE Transactions on Systems, Man, and Cybernetics, Part B (Cybernetics)*, 34(4):1630–1645, 2004.

[45] Ramón González, Francisco Rodríguez, and José Luis Guzmán. *Autonomous Tracked Robots in Planar Off-Road Conditions: Modelling, Localization, and Motion Control.* Springer International Publishing, 2014.

[46] R. Gourdeau. Object-oriented programming for robotic manipulator simulation. *Robotics Automation Magazine, IEEE*, 4(3):21–29, Sep 1997.

[47] Håvard Fjær Grip, Ali Saberi, and Tor A Johansen. Observers for interconnected nonlinear and linear systems. *Automatica*, 48(7):1339–1346, 2012.

[48] Madan M Gupta, Ivo Bukovsky, Noriyasu Homma, Ashu MG Solo, and Zeng-Guang Hou. Fundamentals of higher order neural networks for. *Network and Communication Technology Innovations for Web and IT Advancement*, page 103, 2012.

[49] W. M. Haddad, V.-S. Chellaboina, J. L. Fausz, and C. Abdallah. Optimal discrete-time control for non-linear cascade systems. *Journal of The Franklin Institute*, 335:827–839, 1998.

[50] M.T. Hagan, H.B. Demuth, M.H. Beale, and O. De Jesús. *Neural Network Design (2nd Edition)*. Martin Hagan, 2014.

[51] Abdullah Aamir Hayat, Ratan O. M. Sadanand, and Subir. K. Saha. Robot manipulation through inverse kinematics. In *Proceedings of the 2015 Conference on Advances In Robotics*, AIR '15, pages 48:1–48:6, New York, NY, USA, 2015. ACM.

[52] S. Haykin. *Kalman Filtering and Neural Networks*. John Wiley and Sons, New York, NY, USA, 2001.

[53] S. Haykin. *Kalman Filtering and Neural Networks*. Wiley, 2004.

[54] S.S. Haykin. *Neural Networks and Learning Machines*. Prentice Hall, 2009.

[55] Dietmar Hildenbrand. *Foundations of Geometric Algebra Computing*, volume 8. Springer, 2012.

[56] M. Huichao, L. Shurong, and C. Haiyang. Robust backstepping tracking control for mobile robots. In *Proceedings of the 31st Chinese Control Conference*, pages 4842–4846, July 2012.

[57] S. Hutchinson, G.D. Hager, and P.I Corke. A tutorial on visual servo control. *Robotics and Automation, IEEE Transactions on*, 12(5):651–670, Oct 1996.

[58] A. Iftar. Decentralized optimal control with overlapping decompositions. In *Systems Engineering, 1991., IEEE International Conference on*, pages 299–302, Dayton, OH, USA, Aug 1991.

[59] P. A. Ioannou and J. Sun. *Robust Adaptive Control*. Prentice Hall, Inc, New Jersey, USA, 1996.

[60] Yu Jiang and Zhong-Ping Jiang. Robust adaptive dynamic programming for large-scale systems with an application to multimachine power systems. *Circuits and Systems II: Express Briefs, IEEE Transactions on*, 59(10):693–697, 2012.

[61] Christopher John Cornish, Hellaby Watkins, and Dayan Peter. Q-learning. *Machine Learning*, 8(3-4):279–292, 1992.

[62] T. Kageyama and K. Ohnishi. An architecture of decentralized control for multi-degrees of freedom parallel manipulator. In *Advanced Motion Control, 2002. 7th International Workshop on*, pages 74–79, Maribor, Slovenia, 2002.

[63] Karanjit Kalsi, Jianming Lian, and Stanislaw H Zak. Decentralized dynamic output feedback control of nonlinear interconnected systems. *Automatic Control, IEEE Transactions on*, 55(8):1964–1970, 2010.

[64] Dongshin Kim, Jie Sun, Sang Min Oh, James M Rehg, and Aaron F Bobick. Traversability classification using unsupervised on-line visual learning for outdoor robot navigation. In *Robotics and Automation, 2006. ICRA 2006. Proceedings 2006 IEEE International Conference on*, pages 518–525. IEEE, 2006.

[65] D. E. Kirk. *Optimal Control Theory: An Introduction*. Dover Publications, Englewood Cliffs, NJ, USA, April 2004.

[66] D.E. Kirk. *Optimal Control Theory: An Introduction*. Dover Publications, 2004.

[67] D.E. Kirk. *Optimal Control Theory: An Introduction*. Dover Books on Electrical Engineering Series. Dover Publications, 2004.

[68] M. Krstic, P. V. Kokotovic, and I. Kanellakopoulos. *Nonlinear and Adaptive Control Design*. John Wiley and Sons, Inc., New York, NY, USA, 1st edition, 1995.

[69] Javad Lavaei. Decentralized implementation of centralized controllers for interconnected systems. *Automatic Control, IEEE Transactions on*, 57(7):1860–1865, 2012.

[70] Yann LeCun and Corinna Cortes. MNIST handwritten digit database. 2010.

[71] F. L. Lewis and V. L. Syrmos. *Optimal Control*. John Wiley and Sons, New York, New York, USA, 1995.

[72] Frank L Lewis, Hongwei Zhang, Kristian Hengster-Movric, and Abhijit Das. *Cooperative control of multi-agent systems: optimal and adaptive design approaches*. Springer Science; Business Media, 2013.

[73] W. Lin and C. I. Byrnes. Design of discrete-time nonlinear control systems via smooth feedback. *Automatic Control, IEEE Transactions on*, 39(11):2340–2346, Nov 1994.

[74] Derong Liu, Ding Wang, and Hongliang Li. Decentralized stabilization for a class of continuous-time nonlinear interconnected systems using online learning optimal control approach. *Neural Networks and Learning Systems, IEEE Transactions on*, 25(2):418–428, 2014.

[75] Victor G. Lopez, Alma Y. Alanis, Edgar N. Sanchez, and Jorge Rivera. Real-time implementation of neural optimal control and state estimation for a linear induction motor. *Neurocomputing*, 152:403 – 412, 2015.

[76] Carlos Lopez-Franco, Javier Gomez-Avila, Alma Y. Alanis, Nancy Arana-Daniel, and Carlos Villaseor. Visual servoing for an autonomous hexarotor using a neural network based pid controller. *Sensors*, 17(8), 2017.

[77] M. Lopez-Franco, A. Salome-Bayln, A. Y. Alanis, and N. Arana-Daniel. Discrete super twisting control algorithm for the nonholonomic mobile robots tracking problem. In *2011 8th International Conference on Electrical Engineering, Computing Science and Automatic Control*, pages 1–5, Oct 2011.

[78] M. Lopez-Franco, A. Salome-Baylon, Alma Y. Alanis, and N. Arana-Daniel. Discrete super twisting control algorithm for the nonholonomic mobile robots tracking problem. In *Electrical Engineering Computing Science and Automatic Control (CCE), 2011 8th International Conference on*, pages 1–5, 2011.

[79] Michel Lopez-Franco, Carlos Lopez-Franco Daniel Landa, Alma Y. Alanis, and Nancy Arana-Daniel. Discrete-Time Inverse Optimal Neural Control for a Tracked All Terrain Robot. In *XVI IEEE Autumn Meeting of Power, Electronics and Computer Science ROPEC 2014 INTERNACIONAL*, pages 70–75, 2014.

[80] Michel Lopez-Franco, Edgar N Sanchez, Alma Y Alanis, Carlos Lopez-Franco, and Nancy Arana-Daniel. Discrete-time decentralized inverse optimal neural control combined with sliding mode for mobile robots. In *World Automation Congress (WAC), 2014*, pages 496–501. IEEE, 2014.

[81] G. Lopez-Gonzalez, N. Arana-Daniel, and E Bayro-Corrochano. Parallel clifford support vector machines using the gaussian kernel. *Adv. Appl. Clifford Algebras*, 2016.

[82] G. Lopez-Gonzalez, N. Arana-Daniel, and E Bayro-Corrochano. Quaternion support vector classifier. *Intelligent Data Analysis*, 20(1):109–119, 2016.

[83] Gehová López-González, Nancy Arana-Daniel, and Eduardo Bayro-Corrochano. Conformal hough transform for 2d and 3d cloud points. In

Iberoamerican Congress on Pattern Recognition, pages 73–83. Springer, 2013.

[84] D. G. Lowe. Object recognition from local scale-invariant features. In *Proceedings of the Seventh IEEE International Conference on Computer Vision*, volume 2, pages 1150–1157 vol.2, 1999.

[85] A. Mahajan. Optimal decentralized control of coupled subsystems with control sharing. In *Decision and Control and European Control Conference (CDC-ECC), 2011 50th IEEE Conference on*, pages 5726–5731, Orlando, FL, USA, Dec 2011.

[86] E. Malis, F. Chaumette, and S. Boudet. 2 1/2 d visual servoing. *Robotics and Automation, IEEE Transactions on*, 15(2):238–250, Apr 1999.

[87] Henri Michel and Patrick Rives. *Singularities in the determination of the situation of a robot effector from the perspective view of 3 points*. PhD thesis, INRIA, 1993.

[88] S. A. A. Moosavian and A. Kalantari. Experimental slip estimation for exact kinematics modeling and control of a Tracked Mobile Robot. In *2008 IEEE/RSJ International Conference on Intelligent Robots and Systems*, pages 95–100, 2008.

[89] Mostafa Moussid, Adil Sayouti, and Hicham Medromi. Dynamic modeling and control of a hexarotor using linear and nonlinear methods. *International Journal of Applied Information Systems*, 9(5), 2015.

[90] Patrick F Muir and Charles P Neuman. Kinematic modeling for feedback control of an omnidirectional wheeled mobile robot. In *Autonomous robot vehicles*, pages 25–31. Springer, 1990.

[91] Liz Murphy and Paul Newman. Risky planning on probabilistic costmaps for path planning in outdoor environments. *IEEE Transactions on Robotics*, 29(2):445–457, 2013.

[92] C. Lopez-Franco N. Arana Daniel, R. Valencia-Murillo and A. Alanis-Garcia. Rough terrain perception through geometric entities for robot navigation. In *In 2nd International Conference on Advances in Computer Science and Engineering, CSE 2013, Advances in Intelligent Systems Research*, pages 309–314, 2013.

[93] D.S. Naidu. *Optimal Control Systems*. Electrical engineering textbook series. CRC Press, 2003.

[94] K.S. Narendra and K. Parthasarathy. Identification and control of dynamical systems using neural networks. *Neural Networks, IEEE Transactions on*, 1(1):4–27, 1990.

[95] M Norgaard. *Neural Networks for Modelling and Control of Dynamic Systems: A Practitioner's Handbook.* Springer London, 2000.

[96] K. Ogata. *Discrete-time Control Systems.* Prentice-Hall International, 1995.

[97] T. Ohsawa, A. M. Bloch, and M. Leok. Discrete Hamilton-Jacobi theory and discrete optimal control. In *Decision and Control (CDC), 2010 49th IEEE Conference on,* pages 5438–5443, Atlanta, GA, USA, Dec 2010.

[98] Fernando Ornelas, Edgar N. Sanchez, and Alexander G. Loukianov. Discrete-time inverse optimal control for nonlinear systems trajectory tracking. In *Decision and Control (CDC), 2010 49th IEEE Conference on,* pages 4813–4818, 2010.

[99] F. Ornelas-Tellez, E. N. Sanchez, R. Garcia-Hernandez, J.A. Ruz-Hernandez, and J.L. Rullan-Lara. Neural inverse optimal control for discrete-time uncertain nonlinear systems stabilization. In *Neural Networks (IJCNN), The 2012 International Joint Conference on,* pages 1–6, June 2012.

[100] B. S. Park, S. J. Yoo, J. B. Park, and Y. H. Choi. A simple adaptive control approach for trajectory tracking of electrically driven nonholonomic mobile robots. *Control Systems Technology, IEEE Transactions on,* 18(5):1199 –1206, sept. 2010.

[101] R. Paul. *Robot Manipulators: Mathematics, Programming and Control.* MIT Press, Cambridge, MA, 1982.

[102] Christian Perwass, Vladimir Banarer, and Gerald Sommer. Spherical decision surfaces using conformal modelling. *Lecture notes in computer science,* pages 9–16, 2003.

[103] Christian B. U. Perwass. *Geometric algebra with applications in engineering,* volume 4 of *Geometry and Computing.* Springer, Berlin; Heidelberg, 2009.

[104] Quanser Qbot. Quanser qbot: User manual, number: 830, revision: 7. https://www.gumstix.com.

[105] G. Quintal, E. N. Sanchez, A. Y. Alanis, and N. G. Arana-Daniel. Real-time FPGA decentralized inverse optimal neural control for a shrimp robot. In *System of Systems Engineering Conference (SoSE), 2015 10th,* pages 250–255, 2015.

[106] Nathan D Ratliff, David Silver, and J Andrew Bagnell. Learning to search: Functional gradient techniques for imitation learning. *Autonomous Robots,* 27(1):25–53, 2009.

[107] John Rebula, Greg Hill, Brian Bonnlander, Matthew Johnson, Peter Neuhaus, Carlos Perez, John Carff, William Howell, and Jerry Pratt. Learning terrain cost maps. In *Robotics and Automation, 2008. ICRA 2008. IEEE International Conference on*, pages 2217–2217. IEEE, 2008.

[108] Jorge D Rios, Alma Y Alanis, Michel Lopez-Franco, Carlos Lopez-Franco, and Nancy Arana-Daniel. Real-time neural identification and inverse optimal control for a tracked robot. *Advances in Mechanical Engineering*, 9(3):1687814017692970, 2017.

[109] J Rivera-Mejía, AG Léon-Rubio, and E Arzabala-Contreras. Pid based on a single artificial neural network algorithm for intelligent sensors. *Journal of applied research and technology*, 10(2):262–282, 2012.

[110] Jorge Rivera-Rovelo and Eduardo Bayro-Corrochano. Segmentation and volume representation based on spheres for non-rigid registration. In *International Workshop on Computer Vision for Biomedical Image Applications*, pages 449–458. Springer, 2005.

[111] Jorge Rivera-Rovelo, Eduardo Bayro-Corrochano, and Ruediger Dillmann. Geometric neural computing for 2d contour and 3d surface reconstruction. In *Geometric Algebra Computing*, pages 191–209. Springer, 2010.

[112] Henry Roncancio, Marcelo Becker, Alberto Broggi, and Stefano Cattani. Traversability analysis using terrain mapping and online-trained terrain type classifier. In *Intelligent Vehicles Symposium Proceedings, 2014 IEEE*, pages 1239–1244. IEEE, 2014.

[113] Bodo Rosenhahn and Gerald Sommer. Pose estimation in conformal geometric algebra part i: The stratification of mathematical spaces. *Journal of Mathematical Imaging and Vision*, 22(1):27–48, 2005.

[114] G. A. Rovithakis and M. A. Chistodoulou. *Adaptive Control with Recurrent High -Order Neural Networks*. London, UK, 2000.

[115] George A Rovithakis and Manolis A Christodoulou. *Adaptive control with recurrent high-order neural networks: theory and industrial applications*. Springer Science & Business Media, 2012.

[116] S. Samarasinghe. *Neural Networks for Applied Sciences and Engineering: From Fundamentals to Complex Pattern Recognition*. CRC Press, 2016.

[117] E. N. Sanchez, A. Y. Alanis, and A. G. Loukianov. *Discrete-time High Order Neural Control*. Springer-Verlag, Berlin, Germany, 2008.

[118] E. N. Sanchez and F. Ornelas-Tellez. *Discrete-Time Inverse Optimal Control for Nonlinear Systems*. CRC Press, Boca Raton, FL, USA, 2013.

[119] Edgar N Sanchez, Alanis Y Alanis, and Alexander G Loukianov. Discrete-time high order neural control. *Berlin, Germany: Springer-Verlag*, 10:978–3, 2008.

[120] E.N. Sanchez, A.Y. Alanís, and A.G. Loukianov. *Discrete-Time High Order Neural Control: Trained with Kalman Filtering*. Springer Berlin Heidelberg, 2008.

[121] E.N. Sanchez and F. Ornelas-Tellez. *Discrete-Time Inverse Optimal Control for Nonlinear Systems*. CRC Press, 2016.

[122] Michael David Schmidt. Simulation and control of a quadrotor unmanned aerial vehicle. 2011.

[123] Qiang Sheng, Zhuang Xianyi, Wang Changhong, XZ Gao, and Liu Zilong. Design and implementation of an adaptive pid controller using single neuron learning algorithm. In *Intelligent Control and Automation, 2002. Proceedings of the 4th World Congress on*, volume 3, pages 2279–2283. IEEE, 2002.

[124] Robert Shorten and Roderick Murray-Smith. On normalising radial basis function networks. In *Proceedings of the Fourth Irish Neural Network Conference, University College Dublin, Ireland*, pages 213–217, 1994.

[125] Huang Shoudong, Wang Zhan, and G. Dissanayake. Time optimal robot motion control in simultaneous localization and map building (slam) problem. In *Intelligent Robots and Systems, 2004. (IROS 2004). Proceedings. 2004 IEEE/RSJ International Conference on*, volume 3, pages 3110–3115 vol.3, 2004.

[126] R. Siegwart, I.R. Nourbakhsh, and D. Scaramuzza. *Introduction to Autonomous Mobile Robots*. MIT Press, 2011.

[127] W.D. Smart and L.P. Kaelbling. Effective reinforcement learning for mobile robots. In *Robotics and Automation, 2002. Proceedings. ICRA '02. IEEE International Conference on*, volume 4, pages 3404–3410 vol.4, 2002.

[128] Boris Sofman, Ellie Lin, J Andrew Bagnell, John Cole, Nicolas Vandapel, and Anthony Stentz. Improving robot navigation through self-supervised online learning. *Journal of Field Robotics*, 23(11-12):1059–1075, 2006.

[129] Y. Song and J.W. Grizzle. The Extended Kalman Filter as a Local Asymptotic Observer for Nonlinear Discrete-Time Systems. In *American Control Conference, 1992*, pages 3365–3369, 1992.

[130] W. Sun, H. Gao, and O. Kaynak. Finite frequency H∞ control for vehicle active suspension systems. *IEEE Transactions on Control Systems Technology*, 19(2):416–422, March 2011.

[131] W. Sun, H. Gao, and O. Kaynak. Vibration isolation for active suspensions with performance constraints and actuator saturation. *IEEE/ASME Transactions on Mechatronics*, 20(2):675–683, April 2015.

[132] Richard S. Sutton and Andrew G. Barto. *Reinforcement Learning: An Introduction*. MIT Press, 1998.

[133] Carlos Villaseñor, Nancy Arana-Daniel, Alma Y Alanís, and Carlos López-Franco. Hyperellipsoidal neuron. In *International Joint Conference on Neural Networks (IJCNN) 2017*, pages 788–794. IEEE, 2017.

[134] P.I. Viñuela and I.M.G. Leon. *Redes de neuronas artificiales: un enfoque practico*. Pearson Educacion - Prentice Hall, 2004.

[135] Ning Wang. A generalized ellipsoidal basis function based online self-constructing fuzzy neural network. *Neural processing letters*, 34(1):13–37, 2011.

[136] Z. P. Wang, W. R. Yang, and G. X. Ding. Sliding Mode Control for Trajectory Tracking of Nonholonomic Wheeled Mobile Robots Based on Neural Dynamic Model. In *2010 Second WRI Global Congress on Intelligent Systems*, volume 2, pages 270–273, 2010.

[137] Hellaby Watkins and Christopher John Cornish. *Learning from Delayed Rewards*. PhD thesis, King's College, Cambridge, UK, 1989.

[138] LEEE Weiss, ARTHURC Sanderson, and CHARLESP Neuman. Dynamic sensor-based control of robots with visual feedback. *IEEE Journal on Robotics and Automation*, 3(5):404–417, 1987.

[139] J. Y. Wong and Wei Huang. Wheels vs. tracks A fundamental evaluation from the traction perspective. *Journal of Terramechanics*, 43(1):27 – 42, 2006.

[140] Rui Xu, Donald Wunsch, et al. Survey of clustering algorithms. *Neural Networks, IEEE Transactions on*, 16(3):645–678, 2005.

[141] Jianhua Yang, Wei Lu, and Wenqi Liu. Pid controller based on the artificial neural network. In *International Symposium on Neural Networks*, pages 144–149. Springer, 2004.

[142] J. Yao, Z. Jiao, and D. Ma. Extended-state-observer-based output feedback nonlinear robust control of hydraulic systems with backstepping. *IEEE Transactions on Industrial Electronics*, 61(11):6285–6293, Nov 2014.

[143] J. Yao, Z. Jiao, D. Ma, and L. Yan. High-accuracy tracking control of hydraulic rotary actuators with modeling uncertainties. *IEEE/ASME Transactions on Mechatronics*, 19(2):633–641, April 2014.

[144] Julio Zamora-Esquivel. G 6, 3 geometric algebra; description and implementation. *Advances in Applied Clifford Algebras*, 24(2):493–514, 2014.

Index